Artificial Intelligence for Computer Games

Pedro Antonio González-Calero
Marco Antonio Gómez-Martín
Editors

Artificial Intelligence for Computer Games

Editors
Pedro Antonio González-Calero
Universidad Complutense de Madrid
Dpto. de Ingeniería del
Software e Inteligencia Artificial
28040, Madrid
Spain
pedro@fdi.ucm.es

Marco Antonio Gómez-Martín
Universidad Complutense de Madrid
Dpto. de Ingeniería del
Software e Inteligencia Artificial
28040, Madrid
Spain
marcoa@fdi.ucm.es

ISBN 978-1-4939-0053-4 ISBN 978-1-4419-8188-2 (eBook)
DOI 10.1007/978-1-4419-8188-2
Springer New York Dordrecht Heidelberg London

Printed on acid-free paper

Springer is part of Springer Science+Business Media (www.springer.com)

Here's to our wives and girlfriends . . . may they never meet!

Groucho Marx

Preface

Games has been an interesting domain for artificial intelligence (AI) research since the origins of the discipline back in the 1950s, although mainly board games such as chess, checkers, or backgammon. Nevertheless, it is between 1999 and 2000 when a number of researches on AI identify interactive video games as an interesting research domain for AI (see, for example, John E. Laird, Michael van Lent: "Human-Level AI's Killer Application: Interactive Computer Games". AAAI/IAAI 2000: 1171–1178). The long-term dream of AI research of building intelligent robots with human-like abilities for interacting in the real world is still far from being fulfilled due, among other causes, to the complexities of sensing and acting in the real world. Video games provide synthetic worlds where complex behavior can be shown but where perception and actuation are perfectly under control and, therefore, became a perfect platform for experimenting with software robots (i.e., agents).

Techniques used for AI in commercial video games are still far from state-of-the art in academia, but with graphics in video games coming close to photo realistic quality, and multi-processor architectures getting common in console and PC game platforms, sophisticated artificial intelligence is getting into the focus of the video game industry as the next big thing for enhancing the player experience, while profiting from the number of spare CPU cycles available in modern hardware. For that reason, industry is getting interested in academic research in AI to provide rich, robust, and scalable techniques for controlling non-player characters and provide richer narrative schemes in games.

This book collects some of the most relevant results from academia in the area of artificial intelligence for games. The selection of contributions has been biased toward rigorous and theoretically grounded work that is also supported with developed prototypes, which should pave the way for the integration of academic AI techniques into state-of-the-art electronic entertainment games. The chapters in the book cover different areas relevant to AI in commercial games: pathfinding, decision making, learning, authoring, and storytelling.

Regarding pathfinding, the book describes recent real-time heuristic search algorithms that alleviate the scalability problem of A* techniques used in commercial games, which at the same time exhibit a visually appealing behavior. Techniques are also presented that, based on the semantic annotation of 3D virtual worlds, can learn pathfinding behavior by analyzing traces from actual players. Tools and techniques are described for incorporating semantic information into the game design and development process, thus improving the embedded information contained in immersive game worlds, and leading to new possibilities for NPC constructions such as meaningful in-game learning and agent portability.

Regarding decision making, the book describes new techniques for authoring tools that facilitate the construction by game designers (typically non-programmers) of behavior controlling software, by reusing patterns or actual cases of past behavior, represented as behavior trees. Using domination games as a test bed, the book describes different approaches for building cooperative agents, with different requirements for knowledge engineering, from purely hand-coded to inductive approaches. Techniques for automatically or semi-automatically learning complex behavior from recorded traces of human players using different combinations of reinforcement learning and case-based reasoning are also described.

Much research on artificial intelligence in games has been devoted to creating opponents that play competently against human players, while an alternative goal is to try to deliver the best possible experience within the context of the game. This novel goal is much more attainable by approaching AI reasoning for games as "storytelling reasoning." Several technological approaches are presented in the context of such a perspective, including the use of planning techniques for camera placement and sequencing of plot points in a game, and constraint optimization for automatically adapting lighting qualities of a scene to the player preferences.

Key results from applied research on AI within the last 10 years have been collected here to provide a reference work for both academia and industry that will help to close the gap between both worlds.

Madrid
October 2010

Pedro Antonio González-Calero
Marco Antonio Gómez-Martín

Contents

Contents

Contributors

Bryan Auslander Department of Computer Science & Engineering, Lehigh University, Bethlehem, PA 18015, USA, bla204@lehigh.edu

Yngvi Björnsson School of Computer Science, Reykjavik University, IS-101 Menntavegur 1, Reykjavik, Iceland, yngvi@ru.is

Vadim Bulitko Department of Computing Science, University of Alberta, Edmonton, AB T6G 2E8, Canada, bulitko@ualberta.ca

Belén Díaz-Agudo Complutense University of Madrid, Madrid, Spain, belend@sip.ucm.es

Priyesh N. Dixit The University of North Carolina at Charlotte, 9201 University City Blvd., Charlotte, NC 28223, USA, pndixit@uncc.edu

Magy Seif El-Nasr Simon Fraser University, Burnaby, BC, Canada V5A 1S6, magy@sfu.ca

Gonzalo Flórez-Puga Complutense University of Madrid, Madrid, Spain, gflorez@fdi.ucm.es

Marco Antonio Gómez-Martín Universidad Complutense de Madrid, Dpto. de Ingeniería del Software e Inteligencia Artificial, 28040, Madrid, Spain, marcoa@fdi.ucm.es

Pedro P. Gómez-Martín Complutense University of Madrid, Madrid, Spain, pedrop@fdi.ucm.es

Pedro Antonio González-Calero Universidad Complutense de Madrid, Dpto. de Ingeniería del Software e Inteligencia Artificial, 28040, Madrid, Spain, pedro@fdi.ucm.es

D. Hunter Hale The University of North Carolina at Charlotte, 9201 University City Blvd., Charlotte, NC 28223, USA, dhhale@uncc.edu

Frederick W.P. Heckel The University of North Carolina at Charlotte, 9201 University City Blvd., Charlotte, NC 28223, USA, fheckel@uncc.edu

Chad Hogg Department of Computer Science & Engineering, Lehigh University, Bethlehem, PA 18015, USA, cmh204@lehigh.edu

Arnav Jhala University of California Santa Cruz, Santa Cruz, CA 95064, USA, jhala@cs.ucsc.edu

Ramon Lawrence Computer Science, University of British Columbia Okanagan, 3333 University Way, Kelowna, BC V1V 1V7, Canada, ramon.lawrence@ubc.ca

Stephen Lee-Urban Department of Computer Science & Engineering, Lehigh University, Bethlehem, PA 18015, USA, sml3@lehigh.edu

David Llansó Complutense University of Madrid, Madrid, Spain, llanso@fdi.ucm.es

Héctor Muñoz-Avila Department of Computer Science & Engineering, Lehigh University, Bethlehem, PA 18015, USA, hem4@lehigh.edu

Santiago Ontañón Artificial Intelligence Research Institute (IIIA-CSIC), Campus UAB, 08193 Bellaterra, Spain, santi@iiia.csic.es

Ashwin Ram CCL, Cognitive Computing Lab, Georgia Institute of Technology, Atlanta, GA 30332, USA, ashwin@cc.gatech.edu

Chinmay Rao Penn State University, University Park, PA 16801, USA, chinmayrao@psu.edu

Mark Riedl Georgia Institute of Technology, Atlanta, GA 30332, USA, riedl@cc.gatech.edu

Megan Smith Department of Computer Science & Engineering, Lehigh University, Bethlehem, PA 18015, USA, mev2@lehigh.edu

Nathan R. Sturtevant Department of Computer Science, University of Denver, Denver, CO 80208, USA, sturtevant@cs.du.edu

David Thue University of Alberta, Edmonton, AB, Canada T6G 2E8, dthue@ualberta.ca

R. Michael Young North Carolina State University, Raleigh, NC 27695, USA, young@cs.ncsu.edu

G. Michael Youngblood The University of North Carolina at Charlotte, 9201 University City Blvd., Charlotte, NC 28223, USA, youngbld@uncc.edu

Joseph Zupko Demiurge Studios, Cambridge, MA 02139, USA, jazupko@jazupko.com

Real-Time Heuristic Search for Pathfinding in Video Games

Vadim Bulitko, Yngvi Björnsson, Nathan R. Sturtevant, and Ramon Lawrence

Abstract Game pathfinding is a challenging problem due to a limited amount of per-frame CPU time commonly shared among many simultaneously pathfinding agents. The challenge is rising with each new generation of games due to progressively larger and more complex environments and larger numbers of agents pathfinding in them. Algorithms based on A* tend to scale poorly as they must compute a complete, possibly abstract, path for each agent before the agent can move. Real-time heuristic search algorithms satisfy a constant bound on the amount of planning per move, independent of problem size. These algorithms are thus a promising approach to large scale multi-agent pathfinding in video games. However, until recently, real-time heuristic search algorithms universally exhibited a visually unappealing "scrubbing" behavior by repeatedly revisiting map locations. This had prevented their adoption by video game developers. In this chapter, we review three modern search algorithms which address the "scrubbing" problem in different ways. Each algorithm presentation is complete with an empirical evaluation on game maps.

1 Introduction and Related Work

Heuristic search is a core area of artificial intelligence (AI) research and its algorithms have been widely used in planning, game-playing, and agent control. In this chapter, we are interested in *real-time* heuristic search algorithms that satisfy a constant upper bound on the amount of planning per action, independent of problem size. This property is important in a number of applications including autonomous robots and agents in video games. A common problem in video games is searching for a path between two locations. In most games, agents are expected to act quickly

V. Bulitko (✉)
Department of Computing Science, University of Alberta, Edmonton, AB T6G 2E8, Canada
e-mail: bulitko@ualberta.ca

P.A. González-Calero and M.A. Gómez-Martín (eds.), *Artificial Intelligence for Computer Games*, DOI 10.1007/978-1-4419-8188-2_1,

in response to player's commands and other agents' actions. As a result, many game companies impose a constant time limit on the amount of path planning per move[1] (e.g., one millisecond for *all* simultaneously moving agents).

While in practice this time limit can be satisfied by limiting problem size a priori, a scientifically more interesting approach is to impose a constant per-action time limit *independent* of the problem size. Doing so severely limits the range of applicable heuristic search algorithms. For instance, static search algorithms such as A* [15], Iterative Deepening A* (IDA*) [24] and PRA* [38,39], re-planning algorithms such as D* [37], anytime algorithms such as ARA* [27], and anytime re-planning algorithms such as AD* [26] cannot guarantee a constant bound on planning time per action. This is because all of them produce a complete, possibly abstract, solution before the first action can be taken. As the problem increases in size, their planning time will inevitably increase, exceeding any a priori finite upper bound.

Real-time search addresses the problem in a fundamentally different way. Instead of computing a complete, possibly abstract, solution before the first action is taken, real-time search algorithms compute (or plan) only a few first actions for the agent to take. This is usually done by conducting a lookahead search of a fixed depth (also known as "search horizon," "search depth," or "lookahead depth") around the agent's current state and using a heuristic (i.e., an estimate of the remaining travel cost) to select the next few actions. The actions are then taken and the planning–execution cycle repeats [25]. Since the goal state is not seen in most such local searches, the agent runs the risks of heading into a dead end or, more generally, selecting suboptimal actions. To address this problem, most real-time heuristic search algorithms update (or learn) their heuristic function over time.

The learning process has precluded real-time heuristic search agents from being widely deployed for pathfinding in video games. The problem is that such agents tend to "scrub" (i.e., repeatedly revisit) the state space due to the need to fill in heuristic depressions [19]. As a result, solution quality can be quite low and, visually, the scrubbing behavior is perceived as irrational.

Since the seminal work on Learning Real-Time A* (LRTA*) [25], researchers have attempted to speed up the learning process. Most of the resulting algorithms can be described by the following four attributes:

The *local search space* is the set of states whose heuristic costs are accessed in the planning stage. The two common choices are full-width limited-depth lookahead [14, 16, 17, 25, 31, 33–36], and A*-shaped lookahead [21, 23]. Additional choices are decision-theoretic-based shaping [32] and dynamic lookahead depth-selection [7, 29]. Finally, searching in a smaller, abstracted state has been used as well [13].

The *local learning space* is the set of states whose heuristic values are updated. Common choices are: the current state only [7, 14, 25, 33–35], all states within the local search space [21, 23], and previously visited states and their neighbors [16, 17, 31, 36].

[1] Henceforth, we will use the terms *action* and *move* synonymously.

A *learning rule* is used to update the heuristic costs of the states in the learning space. The common choices are mini-min [16, 17, 25, 31, 34–36], its weighted versions [33], max of mins [7], modified Dijkstra's algorithm [21], and updates with respect to the shortest path from the current state to the best-looking state on the frontier of the local search space [23]. Additionally, several algorithms learn more than one heuristic function [14, 32, 33].

The *control strategy* decides on the move following the planning and learning phases. Commonly used strategies include: the first move of an optimal path to the most promising frontier state [14, 16, 17, 25], the entire path [7], and backtracking moves [7, 34–36].

Given the multitude of proposed algorithms, unification efforts have been undertaken. In particular, [10] suggested a framework, called Learning Real-Time Search (LRTS), to combine and extend LRTA* [25], weighted LRTA* [33], SLA* [35], SLA*T [34], and, to a large extent, γ-Trap [7].

A breakthrough in performance came with D LRTA* [12] which, for the first time in real-time heuristic search, used automatically selected local subgoals instead of the global goal. The subgoal selection mechanism has later been refined in k Nearest Neighbors LRTA* (kNN LRTA*), which we review in this chapter.

In this chapter, we review the following three modern real-time heuristic search algorithms: kNN LRTA*, TBA*, and RIBS.

kNN LRTA* [8, 9] uses a nearest-neighbor algorithm over a database of solved cases. It introduced the idea of compressing a solution path into a series of subgoals so that each can be "easily" reached from the previous one. In doing so, it uses hill-climbing as a proxy for the notion of "easy reachability by LRTA*."

If precomputing a database of solved cases and compressing them into subgoals are not feasible, then one can use the following two modern real-time heuristic search algorithms.

TBA* [2] is a time-bounded variant of the classic A*. Unlike A* that plans a complete path before committing to the first action, Time-Bounded A* (TBA*) interrupts its planning periodically to act. Because initially a complete path to the goal is unknown, the agent instead moves toward the most promising state on the open list, backtracking its steps as necessary. This interleaving of planning and acting is done in such a way that both real-time behavior and completeness are ensured. Among the attractions of this algorithm are its simplicity and broad applicability as well as the fact that reasonable solution quality and real-time performance is achieved without the need for precomputations or state-space abstractions.

RIBS [40] takes a different approach to learning real-time search. Instead of learning a heuristic estimate of the distance from an arbitrary state to the goal as most algorithms have traditionally done, Real-Time Iterative-Deepening Best-First Search (RIBS) learns accurate distances from the start state. This approach has just recently been explored, and more work is required to deploy this algorithm in commercial games. But, the study of RIBS has lead to critical insights in the performance of real-time algorithms and approaches that are likely to be successful.

The rest of the chapter is organized as follows. In Sect. 2 we formulate the problem. Section 3 presents three classic algorithms that serve as the core to TBA*, kNN LRTA*, and RIBS which are reviewed in Sects. 5–7, respectively. Finally, we discuss applications beyond pathfinding in Sect. 8 and conclude the chapter.

2 Problem Formulation

We define a heuristic search problem as an undirected graph containing a finite set of states (vertices) and weighted edges, with a single state designated as the *goal state*. At every time step, a search agent has a single *current state*, a vertex in the search graph, and takes an *action* (or makes a *move*) by traversing an out-edge of the current state. By traversing an edge between states s_1 and s_2, the agent changes its current state from s_1 to s_2. We say that a state is *visited* by the agent if and only if it is the agent's current state at some point of time. As it is usual in the field of real-time heuristic search, we assume that path planning happens *between* the moves (i.e., the agent does not think while traversing an edge). The "plan a move" – "travel an edge" loop continues until the agent arrives at its goal state, thereby solving the problem.

Each edge has a positive cost associated with it. The total cost of edges traversed by an agent from its start state until it arrives at the goal state is called the *solution cost*. We require algorithms to be *complete* (i.e., produce a path from start to goal in a finite amount of time if such a path exists). In order to guarantee completeness for real-time heuristic search, we make the assumption of safe explorability of our search problems. Specifically, all edge costs are finite and for any states s_1, s_2, s_3, if there is a path between s_1 and s_2 and there is a path between s_1 and s_3, then there is also a path between s_2 and s_3.

Formally, all algorithms discussed in this chapter are applicable to any such heuristic search problem. To keep the presentation focused and intuitive, we use a particular type of heuristic search problems, video game pathfinding in grid worlds, for the rest of the chapter. In video game map settings, states are vacant square grid cells. Each cell is connected to four cardinally (i.e., west, north, east, south) and four diagonally neighboring cells. Outbound edges of a vertex are moves available in the corresponding cell, and in the rest of the chapter we will use the terms *action* and *move* interchangeably. The edge costs are *defined* as 1 for cardinal moves and 1.4 for diagonal moves.[2]

An agent plans its next action by considering states in a local search space surrounding its current position. A *heuristic function* (or simply *heuristic*) estimates the (remaining) travel cost between a state and the goal. It is used by the agent to rank available actions and select the most promising one. Furthermore, we consider only *admissible* and *consistent* heuristic functions which do not overestimate the actual

[2] We use 1.4 instead of the Euclidean $\sqrt{2}$ to avoid errors in floating point computations.

remaining cost to the goal and whose difference in values for any two states does not exceed the cost of an optimal path between these states. In this chapter, we use *octile distance* – the minimum cumulative edge cost between two vertices ignoring map obstacles – as our heuristic. This heuristic is admissible and consistent. An agent can modify its heuristic function in any state to avoid getting stuck in local minima of the heuristic function, as well as to improve its action selection with experience.

We evaluate the algorithms presented in this chapter with respect to several performance measures. First, we measure mean *planning time* in terms of both the number of states expanded[3] as well as the CPU time.

The second performance measure of our study is *sub-optimality* defined as the ratio of the solution cost found by the agent to the minimum solution cost -1 and times 100%. To illustrate, suboptimality of 0% indicates an optimal path and suboptimality of 50% indicates a path 1.5 times as costly as the optimal path. We also measure the precomputation time for kNN LRTA* as well as the memory requirements of all three algorithms.

3 The Core Algorithms

TBA*, RIBS, and kNN LRTA* presented later in this chapter build on three classic heuristic search algorithms: A* [15], IDA* [24], and LRTA* [25]. We briefly review these algorithms and discuss their drawbacks for real-time heuristic search below.

*3.1 A**

The classic A* algorithm [15] is a fundamental algorithm for pathfinding. Given a start state s and a goal state g, it finds a least-cost path between the two states. It is a best-first search algorithm and uses a distance-plus-cost-estimate function to determine which state to expand next. The cost function, denoted $f(n)$, consists of two parts: $f(n) = g(s,n) + h(n,g)$ where $g(s,n)$ is the distance of the shortest path found so far between the start state s and state n, and $h(n,g)$ is the heuristic estimate of the distance cost of traveling from state n to the goal g. The algorithm uses two containers to keep track of its search progress: the *open list* storing states that have been encountered but not expanded yet, and the *closed list* storing states already expanded. The algorithm iteratively picks the state from the open list with the lowest f-cost, expands the state, and places its children on the open list. To determine whether a child state goes into the open list, it cannot already be on the closed list or on the open list with a lower cost. The state just expanded is moved to the closed list. The role of the closed list is both to avoid state re-expansions and

[3] A state is called expanded if all of its immediate children are generated.

to reconstruct the solution path once the goal is found. This continues until the goal state is removed from the open list, in which case the solution path is reconstructed from the closed list.

The algorithm is complete, finds an optimal solution when used with an admissible heuristic, and never re-expands states given a consistent heuristic.

3.2 *Iterative Deepening A**

Early researchers noticed that A* could not solve large problems because it would run out of memory. IDA* [24] was thus developed as an alternate algorithm that could find optimal solutions, like A*, but that would only require memory usage linear in the cost of the solution. Most combinatorial puzzles, which were the original focus of IDA*, have state spaces exponential in the solution cost, and so are a natural fit for IDA*. Henceforth, we will call such problems *exponential domains*.

One way to understand how IDA* works is to contrast it to how A* works. Given a consistent heuristic, the lowest f-cost of any state in A*'s open list will monotonically increase during search. Imagine that we grouped states according to their cost when expanded by A*. For instance, all the states with cost 12 might be expanded first, followed by the states with cost 14, and so on. We demonstrate this in Fig. 1, showing contours that delineate states of each successive cost.

IDA* will first expand these groups of states in the same order as A* (modulo tie-breaking among states with equal f-cost) but will subsequently revisit states in subsequent iterations of the algorithm. It does this because it does not maintain an open list. Instead, it performs multiple depth-first searches, with each search bounded by the best f-cost which has yet to be explored. All the states of a particular cost are explored before the next iteration begins anew. In exponential domains such as common combinatorial puzzles, the largest number of states will be expanded in the last iteration, amortizing away the cost of the previous iterations. Because it can be expensive to maintain an open list, IDA* can be faster than A* in practice.

IDA* works best in exponential domains where the state space does not contain many cycles. It might, therefore, seem that IDA* is not well suited to grid-based worlds. These domains are usually *polynomial*, as the number of states in a map

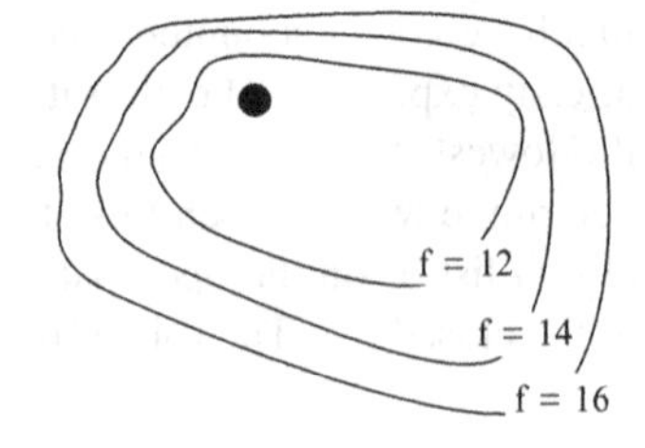

Fig. 1 Iterative Deepening A* (IDA*) search contours: IDA* performs multiple depth-first searches within each successive cost frontier found during search

grows as a polynomial function of length and/or width of the map. Additionally, there are many cycles on grid-based maps. Surprisingly, IDA* can indeed be adapted to perform real-time heuristic search in such domains, as we show below.

3.3 Learning Real-Time A*

The core of most real-time heuristic search algorithms is an algorithm called LRTA* [25]. It is shown below as Algorithm 1 and operates as follows. As long as the goal state $s_{\text{global goal}}$ is not reached, the algorithm interleaves planning and execution in lines 4–7. The original algorithm always sets its goal s_{goal} to be the global goal $s_{\text{global goal}}$. Our generalized version used in this chapter selects s_{goal} dynamically as we detail below. In line 4, a cost-limited breadth-first search with duplicate detection is used to find frontier states with cost up to $g_{\max}$ away from the current state s. For each frontier state $\hat{s}$, its value is the sum of the cost of a shortest path from s to $\hat{s}$, denoted by $g(s,\hat{s})$, and the estimated cost of a shortest path from $\hat{s}$ to s_{goal} [i.e., the heuristic cost $h(\hat{s},s_{\text{goal}})$]. The state that minimizes the sum is identified as s' in line 5. Ties are broken in favor of higher g costs.[4] Remaining ties are broken in a fixed order. The heuristic value of the current state s is updated in line 6 (we keep separate heuristic tables for the different goals and we never decrease heuristics). Finally, we take one step toward the most promising frontier state s' in line 7.

LRTA* is a special case of value iteration or real-time dynamic programming [1] and has a problem that has prevented its use in video game pathfinding. Specifically, it updates a single heuristic value per move on the basis of heuristic values of nearby states. This means that when the initial heuristic values are overly optimistic (i.e., too low), LRTA* will frequently revisit these states multiple times, each time making updates of a small magnitude. This behavior is known as "scrubbing" and appears highly irrational to an observer. Scrubbing is common in pathfinding due to dead ends and corners.

Algorithm 1 LRTA*($s_{\text{start}}, s_{\text{global goal}}, g_{\max}$)

1: $s \leftarrow s_{\text{start}}$
2: **while** $s \neq s_{\text{global goal}}$ **do**
3: **if** no subgoal is selected or the current subgoal is reached **then** select a (new) subgoal s_{goal}
4: generate successor states of s up to $g_{\max}$ cost, generating a frontier
5: find a frontier state s' with the lowest $g(s,s') + h(s',s_{\text{goal}})$
6: update $h(s,s_{\text{goal}})$ to $g(s,s') + h(s',s_{\text{goal}})$
7: change s one step towards s'
8: **end while**

[4] In the rest of the chapter, we use the terms *cost* and *value* interchangeably whenever we refer to f and g functions on states.

There are two fundamental approaches to address problems related to heuristic inaccuracies. First, one can use a more accurate heuristic. Second, one can increase the depth of the lookahead (i.e., by increasing the $g_{\max}$ parameter in LRTA*) to compensate for heuristic inaccuracies. Deeper lookaheads have been generally found beneficial in real-time heuristic search [25], though lookahead pathologies (i.e., detrimental effects of deeper lookahead on solution optimality) have been observed as well [6, 11, 28, 29].

kNN LRTA* takes the former approach and effectively improves the heuristic quality by computing it with respect to a nearby subgoal as opposed to a distant global goal. This is done in an automated fashion as presented below.

4 The Three Modern Algorithms

The three real-time search algorithms discussed in this chapter use A*, IDA*, or LRTA* as their core and enhance them in a number of ways. We review them below. The space constraints preclude us from presenting technical details. Thus, we will focus on the key ideas, the underlying intuition, and support them with highlights of empirical evaluation. We refer the reader to the original publications for additional details [2, 9, 40].

5 Time-Bounded A*

In the absence of precomputed information for guiding the search, LRTA*-like algorithms tend to preform poorly, often revisiting and re-expanding the same states over and over again. In contrast, A* with a consistent heuristic never re-expands a state. However, in A* the first action cannot be taken until an entire solution is planned. As search graphs grow in size, the planning time before the first action will grow, eventually exceeding any fixed cut-off. Consequently, A*-like algorithms violate the real-time property and, thus, do not scale well. One way of alleviating this problem has been to use A* with hierarchies of state-space abstractions, and search first for an approximate path in a highly abstracted state space and then refine it locally in a less abstract one [13, 39]. While faster, their planning time per move still increases with the number of states, making them non-real-time. Another way of addressing the problem is to precompute pathfinding information for expediting the search; however, it may not always be feasible to do so for video game maps. For example, while hours of precomputation per map may be acceptable for maps shipping with a game (as the computation is done beforehand at the game studio), the same is unlikely to be the case for user-generated maps. Also, precomputation is less useful for maps that change frequently during game play (e.g., a bridge or a building is blown-up or a new one built).

Below, we describe a time-bounded version of the A* algorithm, called TBA* [2], that achieves true real-time behavior while requiring neither precomputation nor state space abstractions. It is conceptually the most simple of the three modern pathfinding algorithms described in here.

5.1 TBA: Search*

The TBA* algorithm expands states in an A* fashion, away from the original start state, toward the goal until the goal state is expanded. However, unlike A* that plans a complete path before committing to the first action, TBA* interrupts its search periodically after a fixed number of state expansions and acts. If the complete path to the goal has not yet been found, the agent instead moves toward the most promising state on the open list. This interleaving of planning and acting operations ensures real-time behavior. A key aspect of TBA* over LRTA*-based algorithms is that it retains closed and open lists over its planning steps. Thus, on each planning step it does not start planning from scratch, but continues with its open and closed lists from the previous planning step.

The basic idea behind TBA* is depicted in Fig. 2. S is the start and G the goal, the curves represent A* open list after each expansion time-slice, the small solid circles (a), (b), (c) are states on the open lists with the lowest f-value. The dashed lines are the shortest paths to them. The first three steps of the agent are: $S \rightarrow 1 \rightarrow 2 \rightarrow 1$. The agent backtracks on the last step because the path to the most promising state on the outermost frontier, labeled *(c)*, did not go through state 2 where the agent was situated at the time.

The pseudo-code of TBA* is shown as Algorithm 2. The arguments to the algorithm are the *start* and *goal* states, the search problem P, and the per-move search limit R (expressed as the number of states to expand on each step). The algorithm keeps track of the current location of the agent using the variable *loc*. After initializing the agent location as well as several boolean variables that keep track of the algorithm's internal state (lines 1–4), the algorithm divides up the resource limit as

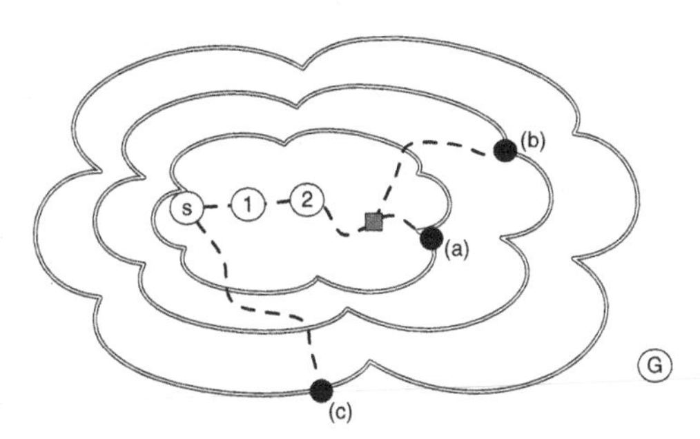

Fig. 2 An example of TBA* in action

Algorithm 2 TBA*($start$, $goal$, P, R)

```
1: loc ← start
2: solutionFound ← false
3: solutionFoundAndTraced ← false
4: doneTrace ← true
5: N_E = ⌊R × r⌋
6: N_T = (R − N_E) × c
7: while loc ≠ goal do
8:    { PLANNING PHASE }
9:    if not solutionFound then
10:      solutionFound ← A*(lists, start, goal, P, N_E)
11:   end if
12:   if not solutionFoundAndTraced then
13:      if doneTrace then
14:         pathNew ← lists.mostPromisingState()
15:      end if
16:      doneTrace ← traceBack(pathNew, loc, N_T)
17:      if doneTrace then
18:         pathFollow ← pathNew
19:         if pathFollow.back() = goal then
20:            solutionFoundAndTraced ← true
21:         end if
22:      end if
23:   end if
24:   { EXECUTION PHASE }
25:   if pathFollow.contains(loc) then
26:      loc ← pathFollow.popFront()
27:   else
28:      if loc ≠ start then
29:         loc ← lists.stepBack(loc)
30:      else
31:         loc ← loc_last
32:      end if
33:   end if
34:   loc_last ← loc
35:   move agent to loc
36: end while
```

it must be shared between state expansion and backtracing[5] operations (lines 5–6). The constants $r \in [0,1]$ and c stand for the fraction of the resource limit to use for state expansions and the relative cost of a expansion compared to backtracing (e.g., a value of 10 indicates that one state expansion takes ten times more time to execute than a backtracing step), respectively. The algorithm then enters the main loop where it repeatedly interleaves *planning* (lines 8–23) and *execution* (lines 24–35) until the agent reaches the goal.

[5] We use the term *backtracing* for the act of tracing a path backwards in the planning phase. The term *backtracking* is used in its usual sense – physically moving (backwards) along the path in the execution phase.

The planning phase proceeds in two steps: first, a fixed number (N_E) of A* state expansions are done (lines 9–11), where both the search problem P and the open/-closed lists are passed to A* as arguments. The former is needed for A* to generate neighbor states and the latter for A* to resume the search from where it left off previously. Second, a new path to follow, *pathNew*, is generated by backtracing the steps from the most promising state on the open list back to the start state. This is done with A* closed list contained in the variable *lists* which also stores A* open list thereby allowing us to run A* in a time-sliced fashion. The function *traceBack* (line 16) backtraces until reaching either the current location of the agent, *loc*, or the *start* state. This is also done in a time-sliced manner (i.e., no more than N_T trace steps per action) to ensure real-time performance. Thus, the backtracing process can potentially span several action steps. Each subsequent call to the *traceBack* routine continues to build the backtrace from the front location of the path passed as an argument and adds the new locations to the front of that path [to start tracing a new path one simply resets the path passed to the routine (lines 13–15)]. Only when the path has been fully traced back, it is set to become the new path for the agent to follow (line 18); until then the agent continues to follow its current path, *pathFollow*.

In the execution phase, the agent does one of the two things as follows. If the agent is already on the path to follow it simply moves one step forward along the path, removing its current location from the path (line 26).[6] On the other hand, if the agent is not on the path – for example, if a different new path has become more promising – then the agent simply starts backtracking its steps one at a time (line 29). The agent will sooner or later step onto the path that it is expected to follow, in the worst case this will happen in the *start* state.

Note that one special case must be handled. Assume a very long new path is being traced back. In general, this causes no problems for the agent as it simply continues to follow its current path until it reaches the end of that path, and if still waiting for the tracing to finish, it simply backtracks toward the *start* state. It is possible, although unlikely, that the agent reaches the start state before a new path becomes available, thus having no path to follow. However, as the agent must act, it simply moves back to the state it came from (line 31).

5.2 TBA*: Properties

Real-time property. The number of state expansions and backtraces performed for each action step is bounded. This is sufficient to claim real-time behavior *provided* that the time it takes to expand or backtrace each state is constant-bounded.

[6] It is not necessary to keep the part of the path already traversed since it can be recovered from the closed list.

In TBA* the open and closed lists grow between action steps; so subsequent planning steps work with larger lists. However, a careful choice of data structures still enables (amortized) constant-time operation.[7]

Completeness. The algorithm expands states in the same manner as A* and is thus guaranteed to find a path from the start state to the goal provided that one exists. The algorithm does additionally guarantee that the agent will get on this solution path and subsequently follow it to the goal. This is done by having the agent backtrack toward the start state when it has no path to follow; during the backtracking process the agent is guaranteed to walk onto the solution path A* found – in the worst case this will be at the start state. TBA* is thus complete.

Memory Complexity. The algorithm uses the same state-expansion strategy as A* and consequently shares the same memory complexity: in the worst case the open and closed lists will cover the entire state space. Traditional heuristic updating real-time search algorithms face a similar worst-case scenario as they may end up having to store an updated heuristic for every state of the search graph. One advantage TBA* has over precomputation-based algorithms is that no memory is allocated for the precomputed data.

5.3 TBA*: Empirical Evaluation

The experiments performed in this section were run using three different maps modeled after game worlds from the popular real-time strategy game Warcraft 3 (shown in Fig. 3). The maps were scaled up to 512×512 cells to increase the problem difficulty [12, 39]. On each map 100 different searches were performed with start and goal locations chosen randomly, although constrained such that the optimal solution cost was between 230 and 320. Each data point we report below is thus an average of 300 different pathfinding problems (3 maps $\times$ 100 searches on each).

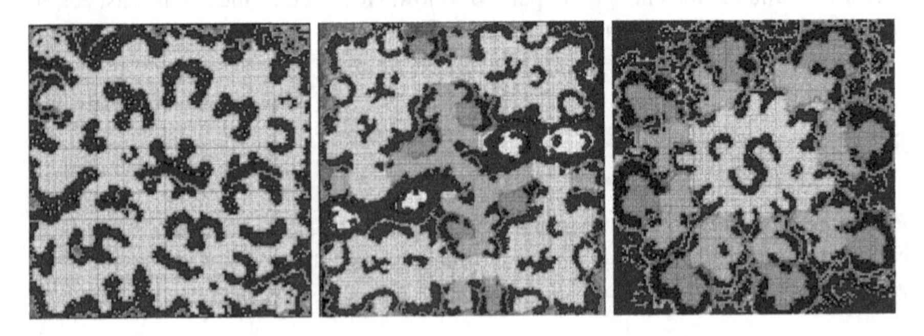

Fig. 3 The maps used in the TBA* experiments

[7] Using the standard heap-based implementation of the open list gives times per move sub-polynomial (logarithmic) in the number of states and, therefore, violates the real-time constraint.

In the experiments that follow, TBA* was matched against two recent real-time search algorithms that have been shown particularly effective in pathfinding on video game maps. They both use state abstraction and precomputation to improve performance. The algorithms and their parameter settings are:

- *PR LRTA** is Path Refinement Learning Real-Time Search [13]. The algorithm runs LRTA* with a fixed search depth d in an abstract space (abstraction level ℓ in a clique abstraction hierarchy [39]) and refines the first action using a corridor-constrained A* running on the original ground-level map. The control parameters are as follows: abstraction level $\ell \in \{3,4,\ldots,7\}$, LRTA* lookahead depth $d \in \{1,3,5,10,15\}$, and LRTA* heuristic weight $\gamma \in \{0.2,0.4,0.6,1.0\}$.
- *D LRTA** is a variant of LRTA* equipped with dynamic search depth and intermediate goal selection [12]. For each map, optimal search depths and intermediate goals (or waypoints) were precomputed beforehand and stored in pattern databases. State abstraction was used to reduce the amount of precomputation. We used the abstraction level of 3 (higher levels of abstraction exceeded the real-time computation cut-off threshold of 1,000 states per action).
- *TBA** is our Time-Bounded TBA*; the resource limit R took on the values $\{10,25,50,75,100,500,1,000\}$, but the values of r and c were fixed at 0.9 and 10, respectively.

A later section of the chapter contrasts the performance of TBA* and kNN LRTA* (thus not included here).

Figure 4 presents the results. The run-time efficiency of the algorithms is plotted. The x-axis represents the amount of work done in terms of the mean number of states expanded per action, whereas the y-axis shows the quality of the solution found relative to an optimal solution (e.g., a value of four indicates that a solution path four times longer than optimal was found). Each point in the figure represents a run of one algorithm with a fixed parameter setting. The closer a point is to the origin, the better performance it represents. Note that we imposed a constraint on the

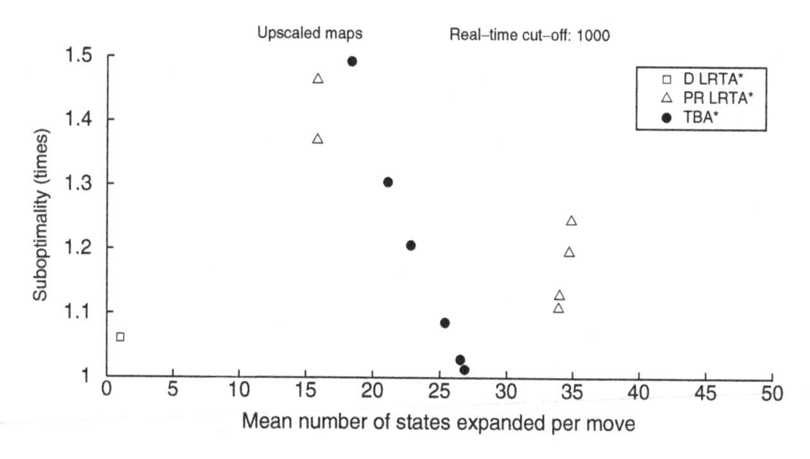

Fig. 4 TBA* compared to other real-time algorithms

parameterization: if the worst-case number of states expanded per action exceeded a cut-off of 1,000 states, then the particular parameter setting was excluded from consideration. Also, to focus on the high-performance area close to the center of origin, we limited the axis limits and, as a result, displayed only a subset of the aforementioned PR LRTA* and D LRTA* parameter combinations.

We see that TBA* performs on par with these algorithms. However, unlike these, it requires neither state-space abstractions nor precomputed pattern databases. This has the advantages of making it both much simpler to implement and better poised for application in nonstationary search spaces, a common condition in video game map pathfinding where other agents or newly constructed buildings can block a path. For example, the data point that is provided for D LRTA*, although showing a somewhat better computation versus suboptimality tradeoff than TBA*, is at the expense of extensive precomputation that can take hours for even a single map.

6 k Nearest Neighbors LRTA*

If an agent is expected to solve a number of problems on the same search graph, then it can make sense to analyze the graph and precompute certain information before attempting to solve the first problem. In the following, we describe one such type of precomputation used in k Nearest Neighbors LRTA* (kNN LRTA*).

6.1 kNN LRTA: Off-Line Subgoal Precomputation*

It has been observed in the literature that common heuristic functions are not uniformly inaccurate [30]. Namely, they tend to be more accurately closer to the goal state and less accurate farther away. The intuition for this fact is as follows: heuristic functions usually ignore certain constraints of the search space. For instance, the Manhattan distance heuristic in a sliding tile puzzle would be perfectly accurate if the tiles could pass through each other. Likewise, the octile distance on a map ignores obstacles. The closer a state is to a goal, the fewer constraints a heuristic function is likely to ignore and, as a result, the more accurate (i.e., closer to the optimal solution cost) the heuristic is likely to be.

We can use these observations to select subgoals dynamically. The idea is straightforward: if being far from the goal leads to grossly inaccurate heuristic values, let us move the goal closer to the agent, thereby improving heuristic accuracy. We can do this by computing the heuristic function with respect to an intermediate, and thus nearby, goal as opposed to a distant global goal – the final destination of an agent. Since an intermediate goal is closer than the global goal, the heuristic values of states around an agent will likely be more accurate. Once the agent gets to an intermediate goal, the next intermediate goal is selected so that the agent makes progress toward its actual global goal. Such dynamic subgoal selection can be carried out using a precomputed subgoal database as described below.

Intuitively, if an LRTA*-controlled agent is in the state s going to the state s_{goal}, then the best subgoal is the state $s_{\text{ideal subgoal}}$ that resides on an optimal path between s and s_{goal} and can be reached by LRTA* along an optimal path with no state revisitation. Given that there can be multiple optimal paths between two states, it is unclear how to computationally efficiently detect the LRTA* agent's deviation from an optimal path *immediately after it occurs*.

On the positive side, detecting state revisitation can be done computationally efficiently by running a simple greedy hill-climbing agent.[8] This is based on the fact that if a hill-climbing agent can reach a state b from a state a without encountering a local minimum or a plateau in the heuristic, then an LRTA* agent will travel from a to b without state revisitation. Thus, we propose an efficiently computable approximation to $s_{\text{ideal subgoal}}$. Namely, we define the subgoal for a pair of states s and s_{goal} as the state $s_{\text{kNN LRTA* subgoal}}$ farthest along an optimal path between s and s_{goal} that can be reached by a simple hill-climbing agent. In summary, we select subgoals to eliminate any scrubbing but do not guarantee that the LRTA* agent keeps on an optimal path between the subgoals. In practice, however, only a tiny fraction of our subgoals are reached by the hill-climbing agent suboptimally and even then the suboptimality is negligible.

This approximation to the ideal subgoal allows us to effectively compute a series of subgoals for a given pair of start and goal states. Intuitively, we *compress* an optimal path into a series of key states such that each of them can be reached from its predecessor without scrubbing. The compression allows us to save a large amount of memory without much impact on time-per-move. Indeed, hill-climbing from one of the key states to the next requires inspecting only the immediate neighbors of the current state and selecting one of them greedily.

However, it is still infeasible to compute and then compress an optimal path between *every* two distinct states in the original search space. We solve this problem by compressing only a predetermined fixed number of optimal paths between random states off-line.

We illustrate this intuition with a simple example. Figure 5 shows kNN LRTA* operation off-line. On this map, two random start and goal pairs are selected, and for each pair an optimal path is computed between the start and goal. Then each path is compressed into a series of subgoals such that each of the subgoals can be reached from the previous one via hill-climbing. The path from S_1 to G_1 is compressed into two subgoals and the other path is compressed into a single subgoal.

[8] In each state, such a simple greedy hill-climbing agent moves to the immediate neighbor with the lowest f-cost. It gives up when all children have their h-cost greater than or equal to the h-cost of the agent's current state.

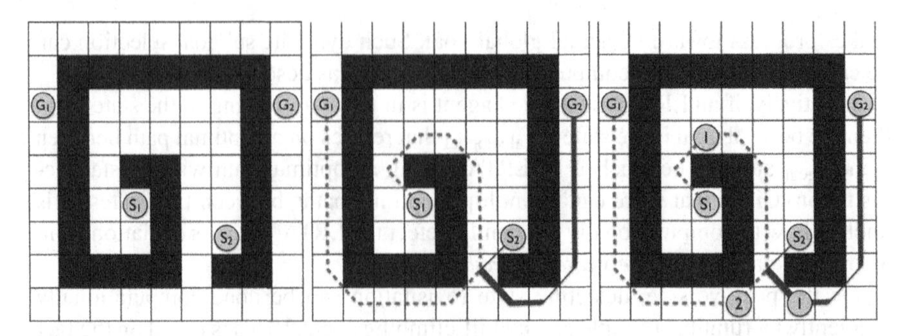

Fig. 5 Example of kNN LRTA* off-line operation. *Left*: two subgoals (start,goal) pairs are chosen: (S_1,G_1) and (S_2,G_2). *Center*: optimal paths between them are computed by running A*. *Right*: the two paths are compressed into a total of three subgoals

6.2 kNN LRTA: Online Search*

Online, kNN LRTA*, tasked with going from s to s_{goal}, retrieves the most similar compressed path from its database and uses the associated subgoals. We define (dis-)similarity of a database path to the agent's current situation as the maximum of the heuristic distances between s and the path's beginning and between s_{goal} and the path's end. We use maximum because we would like *both* ends of the path to be heuristically close to the agent's current state and the goal, respectively. Indeed, the heuristic distance ignores walls and thus a large heuristic distance to the path's either end tends to make that end hill-climbing unreachable.

Note that high similarity (i.e., both distances being low) does not guarantee that the path will be useful to the kNN LRTA* agent. For instance, the beginning of the path can be heuristically very close to the agent but on the other side of a long wall, making it unreachable without a lot of learning and the associated scrubbing. To address this problem, we compliment the fast-to-compute similarity metric with more computationally demanding hill-climbing reachability checks as detailed below.

We illustrate this process by continuing with the simple example introduced in Fig. 5. Once this database of two records is built, kNN LRTA* can be tasked with solving a problem online. In Fig. 6, it is tasked with going from the state S to the state G. The database is scanned and similarity between (S,G) and each of the two database records is determined. The records are sorted by their similarity: (S_1,G_1) followed by (S_2,G_2). Then the agent runs hill-climbing reachability checks[9]: from S to S_i and from G_i to G where i runs the database indices in the order of record similarity. In this example, S_1 is found unreachable by hill-climbing from S and thus the record (S_1,G_1) is discarded. The second record passes hill-climbing checks and the agent is tasked with going to its first subgoal (shown as 1 in the figure).

[9] To satisfy the real-time operation constraint, we set an a priori constant limit on the number of steps in any hill-climbing check online.

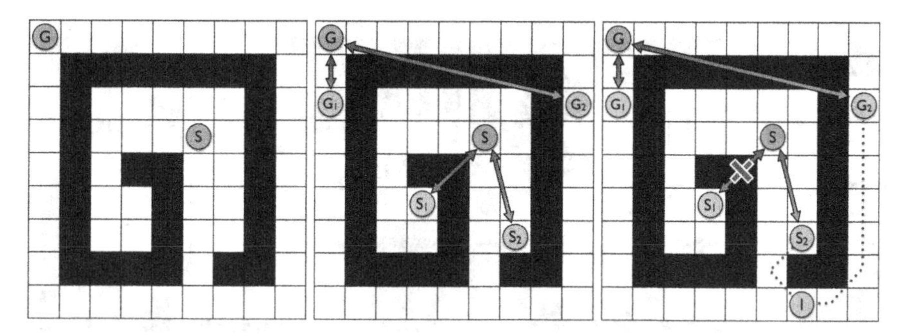

Fig. 6 Example of kNN LRTA* online operation. *Left*: the agent intends to travel from S to G. *Center*: similarity of (S,G) to (S_1,G_1) and (S_2,G_2) is computed. *Right*: while (S_1,G_1) is more similar to (S,G) than (S_2,G_2), its beginning S_1 is not reachable from S via hill-climbing, and hence the record (S_2,G_2) is selected and the agent is tasked with going to subgoal 1

6.3 kNN LRTA: Properties*

Real-time property. On each move kNN LRTA* invokes LRTA*, which expands a constant-bounded number of states. On some moves, kNN LRTA* additionally queries its database to find the appropriate record. Since the database size is independent of the number of states, the query time does not grow with the number of states. The time to sort the records is independent of the total number of states and so are move-limited hill-climbing checks. Therefore, kNN LRTA*'s planning time per move does not grow with the total number of states, satisfying the real-time requirement. Note that in practice larger state spaces tend to require larger databases to provide adequate coverage and maintain solution suboptimality. We study this empirically in Sect. 6.4.

Completeness. Given a problem, the subgoal selection module of kNN LRTA* will either return a database record or instruct LRTA* to go to the global goal. In the latter case, kNN LRTA* is complete because the underlying LRTA* is complete. In the former case, LRTA* is guaranteed to reach the start state of the record due to the way records are picked from the database. LRTA* is then guaranteed to reach the subsequent subgoals due to the completeness of the basic LRTA* and the way the subgoals are constructed.

6.4 kNN LRTA: Empirical Evaluation*

The experiments in this chapter were run on a set of 1,000 randomly generated problems across the four maps shown in Fig. 7. There were 250 problems on each map and they were constrained to have solution cost of at least 1,000. The grid dimensions varied between $4{,}096 \times 4{,}604$ and $7{,}261 \times 4{,}096$ cells. For each problem, we computed an optimal solution cost by running A*. The optimal cost was in

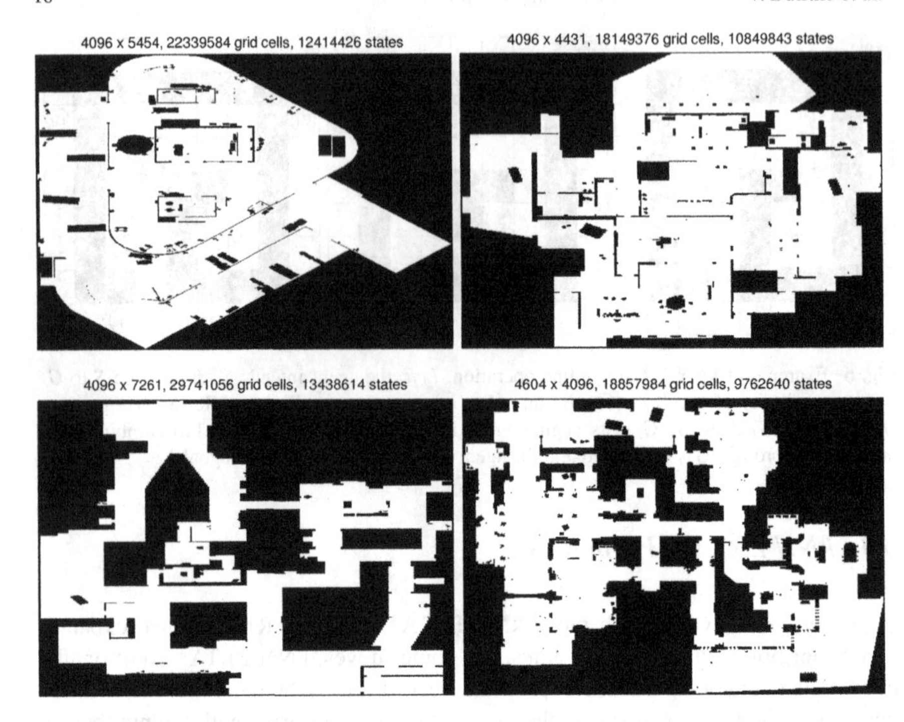

Fig. 7 The maps used in our empirical evaluation

the range of $[1,003.8, 2,999.8]$ with a mean of $1,881.76$, a median of $1,855.2$, and a standard deviation of 549.74. We also measured the A* difficulty defined as the ratio of the number of states expanded by A* to the number of edges in the resulting optimal path. For the $1,000$ problems, the A* difficulty was in the range of $[1, 199.8]$ with a mean of 62.60, a median of 36.47, and a standard deviation of 64.14.

All algorithms compared were implemented in Java using common data structures as much as possible. We used Java version 6 under SUSE Enterprise Linux 10 on a 2.1 GHz AMD Opteron processor with 32 GB of RAM. All timings are reported for single-threaded computations.

We evaluated kNN LRTA* with the following parameters. Database size values were in {1,000, 5,000, 10,000, 40,000, 60,000, 80,000} records. Online, we allowed our hill-climbing test to climb for up to 250 steps before concluding that the destination state is not hill-climbing reachable. This value was picked after some experimentation and had to be appropriate for the record density on the map. To illustrate, a larger database requires fewer hill-climbing steps to maintain the likelihood of finding a hill-climbing reachable record for a given problem.

We ran reachability checks on the 10 most similar records. LRTA*'s parameter $g_{\max}$ was set to the cost of the most expensive edge (i.e., 1.4) so that LRTA* generated only all immediate neighbors of its current state.

We also contrast kNN LRTA*'s performance to that of TBA*, which was run with the time slices of $\{5, 10, 20, 50, 100, 500, 1,000, 2,000, 5,000\}$ and the cost ratio of expanding a state to backtracing set to 10 (explained in the next section).

6.4.1 Solution Suboptimality and Per-Move Planning Time

We begin the comparisons by looking at average solution suboptimality versus average time per move. Table 1 shows the individual values. kNN LRTA* produces the highest quality solutions, followed by TBA*.

TBA* cannot reach kNN LRTA* with the database size of 60,000 and 80,000 records. Additionally, TBA* is noticeably slower per move as it expands more than one state and allocates some time to backtracing as well. The time per move can be decreased by lowering the value of cutoff but already with the cutoff of 10, TBA* produces unacceptably suboptimal solutions (666.5% suboptimal). As a result, kNN LRTA* dominates TBA* by outperforming it with respect to both measures. This is intuitive as TBA* does not benefit from subgoal precomputation.

For the sake of reference, we also included A* results in the table. A* is not a real-time algorithm and its average time per move tends to increase with the number of states in the map. Additionally, it spends most of its time during the first move when it computes the entire path. Subsequent moves require a trivial computation. In the table, we define A*'s mean time per move as the total planning time for a problem divided by the number of moves in the path A* finds. We average this quantity over all problems. kNN LRTA* is about 30 times faster than A* per move.

6.4.2 Database Precomputation Time

Suboptimality versus database precomputation time is shown in Table 2. Note that while the times are roughly between 10 and 100 h, they are reported for single-threaded computations. Because database records are independent of each other,

Table 1 Suboptimality versus time per move

Algorithm	Mean time per move (μs)	Solution suboptimality (%)
kNN LRTA*(10000)	7.56	6,851.62
kNN LRTA*(40000)	6.88	620.63
kNN LRTA*(60000)	6.40	12.77
kNN LRTA*(80000)	6.55	11.96
TBA*(5)	14.31	1,504.54
TBA*(10)	26.34	666.50
TBA*(50)	83.31	131.12
TBA*(100)	117.52	64.66
A*	208.03	0

Table 2 Suboptimality versus database precomputation time

Algorithm	Precomputation time per map (h)	Solution suboptimality (%)
kNN LRTA*(10000)	13.10	6,851.62
kNN LRTA*(40000)	51.89	620.63
kNN LRTA*(60000)	77.30	12.77
kNN LRTA*(80000)	103.09	11.96

the precomputation process scales up linearly with the number of threads. Thus, these times can be decreased by an order of magnitude by simply running the code in parallel on a modern multi-core CPU.

6.4.3 Memory Requirements

Memory is at premium in video games, especially on consoles. TBA* space complexity comes from its open and closed list which it builds online. kNN LRTA* expands only a single state (the agent's current state) and thus has the closed list of one state and the open list of at most eight states (as any grid cell in our maps has at most eight neighbors). However, it consumes memory as it stores updated heuristic values. Additionally, it stores its subgoal databases. We will first focus on the database size. Then we will cover the total memory consumed online: open and closed lists as well as the updated heuristic values.

kNN LRTA* records have two or more states each, and the number of records is fixed by the algorithm parameter. Additionally, kNN LRTA* stores start and end states of each record in a kd-tree. We define *relative database size* as the ratio of the total number of states stored in all records to the total number of map grid cells. The empirical results are found in Table 3.

We will first analyze specifically the amount of memory allocated by the algorithms online. When an algorithm solves a particular problem, we record the maximum size of its open and closed lists as well as the total number of states whose heuristic values were updated. We count each updated heuristic value as one state in terms of storage required.[10] Adding these three measures together, we record the amount of *strictly online memory* per problem. Averaging the strictly online memory over all problems, we list the results in Table 4.

TBA*, as time-sliced A*, does not update heuristic values at all. However, its open and closed lists contribute to the highest memory consumption at $1,353.94$ KB. This is intuitive as TBA* does not use subgoals and therefore must "fill in" potentially large heuristic depressions with its open and closed lists. Also, note that the total size of these lists does not change with the cutoff as state expansions are

Table 3 Database statistics. All values are averages per map. Precomputation time is in hours

Algorithm	Precomputation time	Records	Relative size	Size (MB)
kNN LRTA*(10000)	13.10	10,000	0.00308	0.25
kNN LRTA*(40000)	51.89	40,000	0.01234	1.00
kNN LRTA*(60000)	77.30	60,000	0.01851	1.51
kNN LRTA*(80000)	103.09	80,000	0.02468	2.01

[10] Multiple heuristic updates in the same state do not increase the amount of storage.

Table 4 Strictly online memory versus solution suboptimality

Algorithm	Strictly online memory (KB)	Solution suboptimality (%)
kNN LRTA*(10000)	8.62	6,851.62
kNN LRTA*(40000)	5.04	620.63
kNN LRTA*(60000)	4.23	12.77
kNN LRTA*(80000)	4.22	11.96
TBA*(5)	1,353.94	1,504.54
TBA*(10)	1,353.94	666.50
TBA*(50)	1,353.94	83.31
TBA*(100)	1,353.94	64.66
A*	1,353.94	0

Table 5 Solution suboptimality versus cumulative online memory

Algorithm	Cumulative online memory (KB)	Solution suboptimality (%)
kNN LRTA*(10000)	265.65	6,851.62
kNN LRTA*(40000)	1,034.08	620.63
kNN LRTA*(60000)	1,547.85	12.77
kNN LRTA*(80000)	2,062.20	11.96
TBA*(5)	1,353.94	1,504.54
TBA*(10)	1,353.94	666.50
TBA*(50)	1,353.94	83.31
TBA*(100)	1,353.94	64.66
A*	1,353.94	0

independent of agent's moves in TBA*. A* has identical memory consumption as it expands states in the same way as TBA*. Again, kNN LRTA* dominates TBA* for all cutoff values, using less memory and producing better solutions.

Strictly online memory gives an insight into the algorithms but does not present a complete picture. Specifically, kNN LRTA* must load its databases into its online memory. Thus, we define the *cumulative online memory* as the strictly online memory plus the size of the database loaded. The values are found in Table 5.

TBA* is no longer dominated due to its low memory consumption. The closest comparison is between kNN LRTA* with 60,000 records and TBA*. While kNN LRTA* uses 14% more memory than TBA*, it produces solutions of 1.5–14.2 times better.

7 Real-Time Iterative-Deepening Best-First Search

TBA* is a relatively straightforward extension of A* which allows immediate movement by an agent before A* finds a complete path. However, TBA* is not an agent-centric algorithm. That is, the memory accesses performed by TBA* happen at arbitrary places in the map that may not be local to the agent. This can be

Algorithm 3 RIBS($s_{current}$, g-cost, s_{parent}, s_{goal}) {*Global:* cost_limit ← 0 }

```
1: if s_current is visited the first time then
2:     set parent of s_current to s_parent
3:     store g-cost of s_current
4: end if
5: while s_current is not s_goal do
6:     mark s_current as visited with current cost_limit
7:     for all successor succ_i with f-cost ≤ cost_limit and unvisited with current cost_limit do
8:         if succ_i never expanded or succ_i's parent is s_current then
9:             RIBS(s_current, g-cost + cost(s_current, s_succ_i), s_succ_i, s_goal)
10:        end if
11:    end for
12:    if s_parent is not nil { only occurs at s_start } then
13:        return
14:    else
15:        increase cost_limit
16:    end if
17: end while
```

important if there are cache or memory concerns, where random memory accesses are slow. One way to look at RIBS is that it is an agent-centric version of TBA*; however, there are a few extra pieces that are needed to make RIBS efficient in practice. If an agent-centric approach is not important, TBA* may be a better option.

The basic approach for RIBS is shown in simplified pseudo-code in Algorithm 3. A global cost limit is used as the current estimate of the cost to the goal. An agent begins at the state s_{current} and is passed the last state visited, which is used to set up parent pointers; so the agent can retrace its path if stuck in a dead end (lines 1–4).

Next, the agent computes the f-cost of the successor states, and recursively visits any states which have f-cost less than or equal to the current bound. If all successors are visited without finding the goal, then the agent returns to its parent state. If there is no parent state, then the agent must be in the start state, and there is no path to the goal with the current bound. In this case, the bound is increased and the procedure starts over.

This procedure is essentially the same as IDA*, and so it can be proven that, with a consistent heuristic, when an agent expands a state for the first time it will have discovered an optimal cost path to that state. As a corollary, RIBS is guaranteed to identify an optimal path to the goal state by the time it reaches it. Note that it does not mean that it will have followed such an optimal path. Like other real-time heuristic search agents, a RIBS agent tends to follow suboptimal paths in practice.

A simple agent running RIBS would take one action per move. An agent moves forward on line 9 and moves backwards on line 13. The while statement on line 5 really only serves to keep the agent iterating with increasing cost limits at the start state, as for every other state a parent will be defined causing the while loop to exit at line 13.

This description of the algorithm is fairly simple to implement, but it is missing a few details, such as some of the code for initializing new states, the procedure for updating the cost limit, and some important pruning details. The first two details are relatively straightforward, so we will only discuss the pruning details here. Also note that RIBS is shown as a recursive algorithm that would run until completion, but it is not hard to break this computation into pieces that could be resumed when the time limit for the current action expires.

7.1 RIBS: Intuition

Learning h-values can be slow because inaccurate heuristics values are used to update other (also inaccurate) heuristic values. This is particularly problematic if a learning agent enters a heuristic depression [20], a localized area in the search space where it has to repeatedly revisit states to raise their heuristic values enough to be able to continue to explore other parts of the search space. The more frequent and deeper the depressions are, the more severely the problem manifests itself.

This is illustrated in Fig. 8 with an example of LRTA* behavior on a portion of a map with a local minima in the corner. To simplify the example, diagonal moves have cost 1.5. States are marked with their initial heuristic values. Consider part (a) where the agent is in the shaded state. Using a lookahead of one, the value of the corner heuristic can be updated from 3 to 5, because a neighbor distance 1 has heuristic cost 4. In part (b) the agent moves to the highlighted state and makes a similar update, raising the h-cost to 5.5, before moving to the state updated in part (c), where that state will be updated to have a heuristic value of 5.5.

After three updates, considerable learning still remains. This is because the heuristic is being updated locally from neighboring heuristics, which, due to consistency, cannot be considerably larger. Thus, a state must be visited and updated many times before large changes in the heuristic can occur. As this learning begins far from the goal state, heuristic estimates are likely to be inaccurate. For the same reason that heuristic costs (h-costs) tend to be more accurate closer to the goal state, g-costs tend to be more accurately closer to the start state. RIBS takes advantage of this fact and learns g-costs for all states visited by the agent. Since the agent begins in the start state, such a learning process is more efficient than the h-cost learning of LRTA*: more accurate g costs are learned faster.

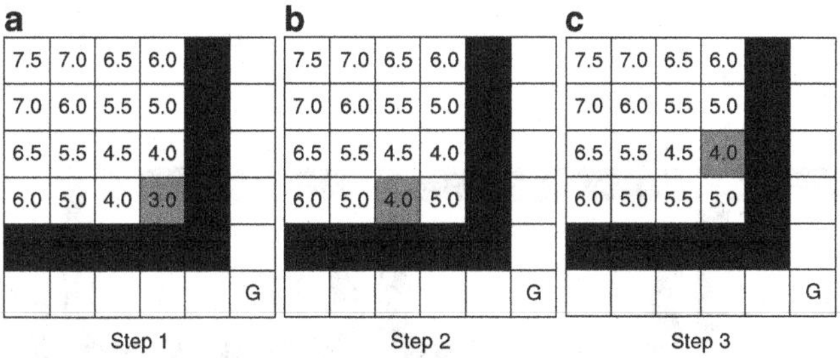

Fig. 8 Learning in a local minima

The second key observation is that (accurate) g-costs can be helpful in escaping heuristic depressions. Not only can they greatly reduce the number of times the agent must revisit states in a heuristic depression, but they are also useful in identifying dead states and redundant paths. Such identifications require accurate costs and, as a result, work much better with g-cost learning than with h-cost learning.

Excluding the start and goal, any state on an optimal path must have neighbors with both higher and lower g-costs. If a state has no neighbors with larger g-costs, then an optimal path to the goal cannot pass through this state. We thus define a *dead state* as follows: Given a start state s and a state n, n is a dead state if n is not the goal state and if for all non-dead neighbors of n, $n_1 \ldots n_i$, $\mathrm{cost}(s,n) \geq \mathrm{cost}(n_i,s)$.

Consider the example in Fig. 9, which shows g-cost estimates for the same problem. Upon reaching the corner, the agent can potentially mark each state with the g-costs in the figure, which are upper bounds on the actual cost to each state. In part (a) of the figure, the agent can see that the highlighted state in the corner is dead, because all neighbors can be reached by shorter paths through other states. After this state is marked dead, in part (b), two more states can be marked dead. Learning that a state is dead only requires visiting a state a single time, unlike learning a heuristic, which may take multiple visits. Fortunately, there is more that can be done if we know the optimal cost to each state.

Consider Fig. 10a. In this case, the states in the corners can be marked as dead and ignored once the optimal cost is discovered. Note, however, that even after removing the dead states there are still many paths that lead out through this room, shown in Fig. 10b. However, because there is only a single doorway to the room, these paths are all redundant. Detecting and ignoring states on such redundant paths offer additional saving. This requires two steps. In Fig. 10c, we show two possible optimal paths leading out of the room. We focus on states A and B, shown in detail in Fig. 10d.

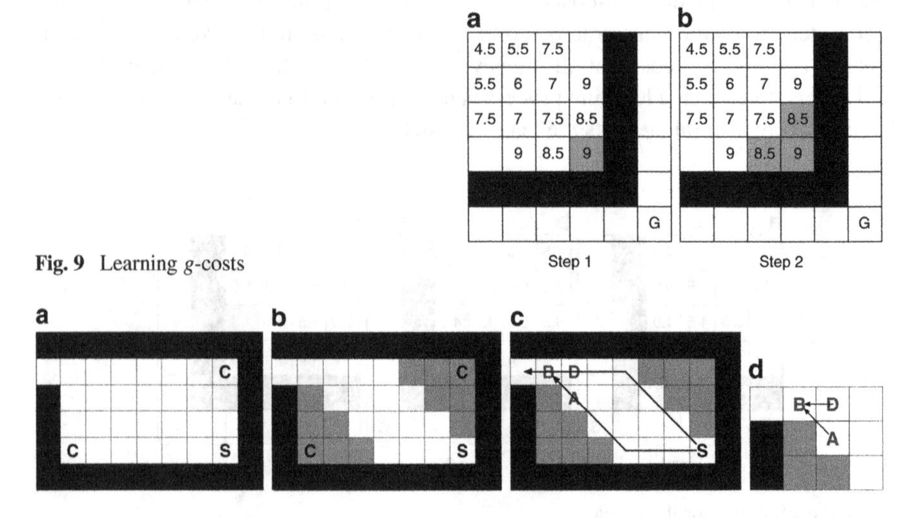

Fig. 9 Learning g-costs

Fig. 10 Marking dead states

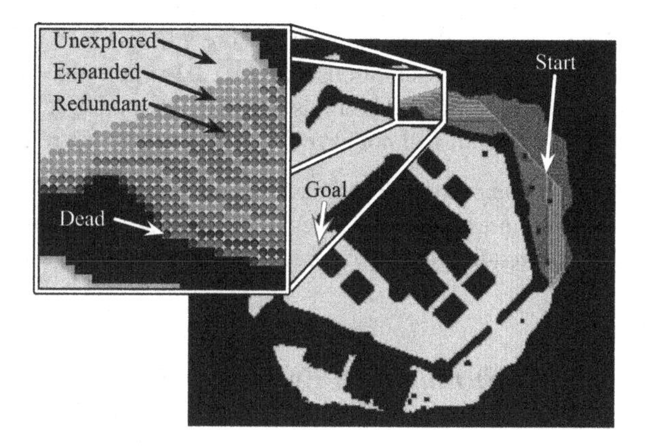

Fig. 11 An example map showing the search in-progress with the different types of states

The first step is to mark all parents which are along optimal paths to a state. Each time a state is generated, if the parent is on an optimal path to the state, the parent is added to the list of optimal parents for that state. In the case of state B, states A and D can both reach B with the same cost, and so B maintains this information.

The second step occurs in the next iteration of search. Suppose that A is visited first. Then, at A we will notice that there is also an optimal path to B through D. Since there are no other optimal paths through A to a successor of A, A can be marked dead or redundant, and B is marked to have a single optimal parent of D. If D were visited first, D would be marked redundant, and only the path through A would be maintained.

In Fig. 11, we show an example of redundant state removal on a fragment of an actual game map. The different types of states are labeled with arrows. Most states have been marked as dead or redundant, meaning that additional exploration focuses just on the successors of a small fringe of previously expanded states.

The effectiveness of dead state and redundant state pruning will depend on the problem being solved. We observe that if previously expanded states can be marked dead and/or redundant at the same rate that new states are expanded, then the performance of RIBS would approach that of TBA*, as TBA* never revisits expanded states. But, RIBS must first mark states as dead and/or redundant to stop visiting them, and searches in multiple iterations, so it cannot completely match the performance of TBA*, especially given that it obeys agent-centric constraints.

7.2 RIBS: Properties

Real-time property. The number of state expansions performed for each step of RIBS can be set to any desired constant. Note that RIBS maintains no open or closed lists and thus does not require sophisticated data representations to satisfy the real-time constraint.

Completeness. RIBS, being at its core a time-sliced version of IDA*, is complete under the same assumptions as IDA*. Also note that, similar to TBA* and unlike kNN LRTA*, when RIBS finds the goal it will have determined the optimal solution path, although it will not have followed that path en route to the goal.

Memory complexity. RIBS has the same worst case as TBA* where it will consider and store information all states in the state space. Unlike kNN LRTA*, however, it does not require loading or precomputing a database.

7.3 RIBS: Empirical Evaluation in Heuristic Depressions

Of all the algorithms discussed in this chapter, RIBS makes the most restrictive assumptions about what an agent running RIBS is able to perform in the environment. RIBS assumes no up-front knowledge of the domain, ruling out any opportunity for precomputation. RIBS also assumes that random access to states far from the agent is not available, ruling out the TBA* approach. TBA* and kNN LRTA* have better performance than RIBS in practice because they do not make such assumptions. As a result, we focus on evaluating RIBS' ability to quickly escape heuristic depressions. We focus on the comparison between g-learning RIBS and h-learning basic LRTA*. This showcases RIBS ability to use its accurately learned g-costs to identify the redundant and dead states as described above.

The basic version of LRTA* which we compare against can be described as follows: The *local search space* only includes the neighbors of the current state. The *local learning space* is only the current state. The *learning rule* is mini-min, and the *control strategy* is to move to the best neighboring state. Although the local search space and learning space are small, increasing their size does not significantly change the results we present here (see [40] for more details).

We experiment on the map in Fig. 12, where the agent starts in the upper left corner and must travel to the lower right corner. The default heuristic leads directly into the corner, from which the agent must then escape. This structure is common in many maps, and so we experiment directly with this example, scaling the size to measure performance.

The results of the comparison are shown in Fig. 13. The x-axis is the number of states in the whole map, while the y-axis is the number of states expanded by

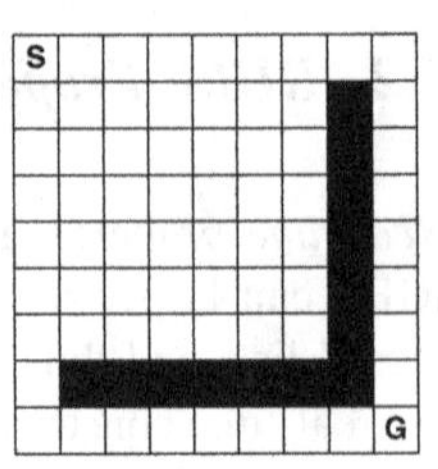

Fig. 12 The example map used to compare RIBS and LRTA* performance

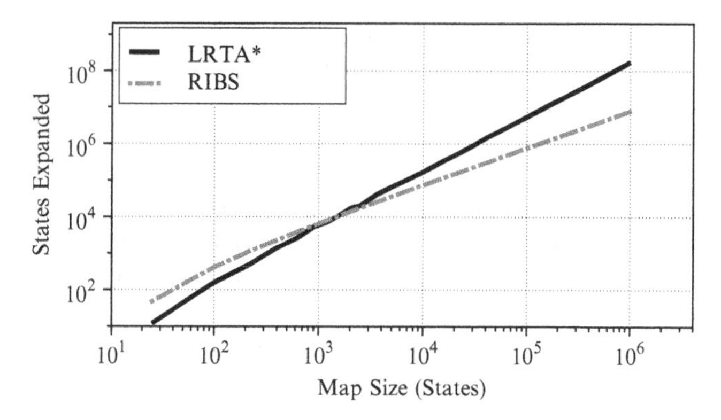

Fig. 13 RIBS versus LRTA* performance on the example map

each algorithm. Note that both axes use a logarithmic scale. The RIBS line approximates $y = 10x$ as the map gets larger, while the LRTA* approximates $y = 0.14x^{1.5}$. This means that, once the map gets large, RIBS can expand each state 10 times before finding the optimal solution. For LRTA*, however, the number of expansions is polynomial in the size of the map. This explains the "scrubbing" behavior of LRTA* – the number of expansions that it takes for LRTA* to escape a local minima can be far more than the number of states in the local minima.

The performance of RIBS and LRTA* crosses when the number of states in the local minima grows to approximately 1,500 states, which corresponds to a 40×40 room in a larger map, a size that is not unrealistic.

8 Future Work

We presented and evaluated kNN LRTA*, TBA*, and RIBS for grid-based pathfinding. Formally, the algorithms are applicable to arbitrary weighted graphs that satisfy the constraints at the beginning of Sect. 2. Thus, in principle, they should be applicable to general planning using the ideas from search-based planners ASP [5], the HSP-family [3], FF [18], SHERPA [22], and LDFS [4]. An actual application is a subject of future work.

The methods can also be further improved and fine tuned in our problem domain of pathfinding in video games. Currently one of the main drawbacks of kNN LRTA* is the long precomputation time needed for generating the off-line databases. While the time is affordable on the game company side, most players would want their home-made game maps to be processed in a matter of seconds or minutes. While the computation can be sped up at a linear scale using multi-core processors, this would still come up short. One of the main focus of future work on kNN LRTA* will thus be to shorten the precomputation time; for example, we might be able to get away with much smaller databases if the database records were generated in a manner such that they produce a better coverage of the state space.

Unlike A*, no real-time algorithm can guarantee finding an optimal path. This is of a little consequence in video game pathfinding as long as approximately optimal paths are found. More importantly is that the agents navigate the game world in a rational way, for example, that they do not show visually jarring or indecisive behavior by frequently changing their mind as of where to go. While both kNN LRTA* and TBA* are much improved in that respect compared to most other mainstream real-time algorithms, such behavior does still occasionally surface, especially in the latter. Even a single incident of an irrational pathfinding behavior can break the player's immersion.

Some preliminary solutions for TBA* are provided in [2] but more effective solutions are required.

RIBS is a promising approach but, given its recency, further empirical evaluation as well as algorithmic improvements are necessary. In particular, variants of RIBS that forgo the eventual identification of optimal paths and, as a result, find better suboptimal solutions can be explored.

9 Conclusions

In this chapter, we considered the problem of real-time heuristic search whose planning time per move does not depend on the number of states. We reviewed three modern algorithms, each with its strengths and weaknesses.

In terms of solution sub-optimality when given equal computing resources (or vice versa, required computing resources for finding equally good solutions), kNN LRTA* shows the best performance. Because pathfinding tends to be a rather computing intensive task in modern games, especially in large game worlds with multiple agents navigating simultaneously, this metric is of an utmost importance. This level of online performance comes at the cost of long off-line precomputation times (hours per map). TBA*, although not being quite as effective as kNN LRTA*, still shows a good performance and has the benefit of not requiring any precomputation. It may thus be better poised for environments that dynamically change during game play. TBA* also uses on average somewhat less memory that kNN LRTA*, which can be an important consideration on some gaming platforms (e.g., consoles). Both algorithms thus appear well poised for video game pathfinding.

RIBS is an interesting way of moving TBA* closer to being an agent-centered approach – an important consideration for some problem domains. The algorithm also provides added insights into how real-time search agents can learn heuristics.

Acknowledgements Parts of this chapter have been previously published as conference and journal articles written by the authors. This research was supported by grants from the National Science and Engineering Research Council of Canada (NSERC) and the Icelandic Centre for Research (RANNÍS). All the research works by Nathan Sturtevant included in this chapter were performed at the University of Alberta. We appreciate the help of Josh Sterling, Stephen Hladky, and Daniel Huntley.

References

1. Barto, A.G., Bradtke, S.J., Singh, S.P.: Learning to act using real-time dynamic programming. Artificial Intelligence **72**(1), 81–138 (1995)
2. Björnsson, Y., Bulitko, V., Sturtevant, N.: TBA*: Time-bounded A*. In: Proceedings of the International Joint Conference on Artificial Intelligence (IJCAI), pp. 431 – 436. AAAI Press, Pasadena, California (2009)
3. Bonet, B., Geffner, H.: Planning as heuristic search. Artificial Intelligence **129**(1–2), 5–33 (2001)
4. Bonet, B., Geffner, H.: Learning depth-first search: A unified approach to heuristic search in deterministic and non-deterministic settings, and its application to MDPs. In: Proceedings of the International Conference on Automated Planning and Scheduling (ICAPS), pp. 142–151. Cumbria, UK (2006)
5. Bonet, B., Loerincs, G., Geffner, H.: A fast and robust action selection mechanism for planning. In: Proceedings of the National Conference on Artificial Intelligence (AAAI), pp. 714–719. AAAI Press / MIT Press, Providence, Rhode Island (1997)
6. Bulitko, V.: Lookahead pathologies and meta-level control in real-time heuristic search. In: Proceedings of the 15th Euromicro Conference on Real-Time Systems, pp. 13–16 (2003)
7. Bulitko, V.: Learning for adaptive real-time search. Tech. Rep. http://arxiv.org/abs/cs.AI/0407016, Computer Science Research Repository (CoRR) (2004)
8. Bulitko, V., Björnsson, Y.: kNN LRTA*: Simple subgoaling for real-time search. In: Proceedings of Artificial Intelligence and Interactive Digital Entertainment (AIIDE), pp. 2–7. AAAI Press, Stanford, California (2009)
9. Bulitko, V., Björnsson, Y., Lawrence, R.: Case-based subgoaling in real-time heuristic search for video game pathfinding. Journal of Artificial Intelligence Research (JAIR) **39**, 269–300 (2010)
10. Bulitko, V., Lee, G.: Learning in real time search: A unifying framework. Journal of Artificial Intelligence Research (JAIR) **25**, 119–157 (2006)
11. Bulitko, V., Li, L., Greiner, R., Levner, I.: Lookahead pathologies for single agent search. In: Proceedings of the International Joint Conference on Artificial Intelligence (IJCAI), pp. 1531–1533. Acapulco, Mexico (2003)
12. Bulitko, V., Luštrek, M., Schaeffer, J., Björnsson, Y., Sigmundarson, S.: Dynamic control in real-time heuristic search. Journal of Artificial Intelligence Research (JAIR) **32**, 419–452 (2008)
13. Bulitko, V., Sturtevant, N., Lu, J., Yau, T.: Graph abstraction in real-time heuristic search. Journal of Artificial Intelligence Research (JAIR) **30**, 51–100 (2007)
14. Furcy, D., Koenig, S.: Speeding up the convergence of real-time search. In: Proceedings of the National Conference on Artificial Intelligence (AAAI), pp. 891–897 (2000)
15. Hart, P., Nilsson, N., Raphael, B.: A formal basis for the heuristic determination of minimum cost paths. IEEE Transactions on Systems Science and Cybernetics **4**(2), 100–107 (1968)
16. Hernández, C., Meseguer, P.: Improving convergence of LRTA*(k). In: Proceedings of the International Joint Conference on Artificial Intelligence (IJCAI), Workshop on Planning and Learning in A Priori Unknown or Dynamic Domains, pp. 69–75. Edinburgh, UK (2005)
17. Hernández, C., Meseguer, P.: LRTA*(k). In: Proceedings of the International Joint Conference on Artificial Intelligence (IJCAI), pp. 1238–1243. Edinburgh, UK (2005)
18. Hoffmann, J.: A heuristic for domain independent planning and its use in an enforced hill-climbing algorithm. In: Proceedings of the 12th International Symposium on Methodologies for Intelligent Systems (ISMIS), pp. 216–227 (2000)
19. Ishida, T.: Moving target search with intelligence. In: National Conference on Artificial Intelligence (AAAI), pp. 525–532 (1992)
20. Ishida, T.: Moving target search with intelligence. In: Proceedings of the National Conference on Artificial Intelligence, pp. 525–532 (1992)
21. Koenig, S.: A comparison of fast search methods for real-time situated agents. In: Proceedings of Int. Joint Conf. on Autonomous Agents and Multiagent Systems, pp. 864–871 (2004)

22. Koenig, S., Furcy, D., Bauer, C.: Heuristic search-based replanning. In: Proceedings of the Int. Conference on Artificial Intelligence Planning and Scheduling, pp. 294–301 (2002)
23. Koenig, S., Likhachev, M.: Real-time adaptive A*. In: Proceedings of the International Joint Conference on Autonomous Agents and Multiagent Systems (AAMAS), pp. 281–288 (2006)
24. Korf, R.: Depth-first iterative deepening: An optimal admissible tree search. Artificial Intelligence **27**(3), 97–109 (1985)
25. Korf, R.: Real-time heuristic search. Artificial Intelligence **42**(2–3), 189–211 (1990)
26. Likhachev, M., Ferguson, D.I., Gordon, G.J., Stentz, A., Thrun, S.: Anytime dynamic A*: An anytime, replanning algorithm. In: ICAPS, pp. 262–271 (2005)
27. Likhachev, M., Gordon, G.J., Thrun, S.: ARA*: Anytime A* with provable bounds on sub-optimality. In: S. Thrun, L. Saul, B. Schölkopf (eds.) Advances in Neural Information Processing Systems 16. MIT, Cambridge, MA (2004)
28. Luštrek, M.: Pathology in single-agent search. In: Proceedings of Information Society Conference, pp. 345–348. Ljubljana, Slovenia (2005)
29. Luštrek, M., Bulitko, V.: Lookahead pathology in real-time path-finding. In: Proceedings of the National Conference on Artificial Intelligence (AAAI), Workshop on Learning For Search, pp. 108–114. Boston, Massachusetts (2006)
30. Pearl, J.: Heuristics. Addison-Wesley, Reading, MA (1984)
31. Rayner, D.C., Davison, K., Bulitko, V., Anderson, K., Lu, J.: Real-time heuristic search with a priority queue. In: Proceedings of the International Joint Conference on Artificial Intelligence (IJCAI), pp. 2372–2377. Hyderabad, India (2007)
32. Russell, S., Wefald, E.: Do the Right Thing: Studies in Limited Rationality. MIT, Cambridge, MA (1991)
33. Shimbo, M., Ishida, T.: Controlling the learning process of real-time heuristic search. Artificial Intelligence **146**(1), 1–41 (2003)
34. Shue, L.Y., Li, S.T., Zamani, R.: An intelligent heuristic algorithm for project scheduling problems. In: Proceedings of the 32nd Annual Meeting of the Decision Sciences Institute. San Francisco (2001)
35. Shue, L.Y., Zamani, R.: An admissible heuristic search algorithm. In: Proceedings of the 7th International Symposium on Methodologies for Intelligent Systems (ISMIS-93), LNAI, vol. 689, pp. 69–75 (1993)
36. Sigmundarson, S., Björnsson, Y.: Value Back-Propagation vs. Backtracking in Real-Time Search. In: Proceedings of the National Conference on Artificial Intelligence (AAAI), Workshop on Learning For Search, pp. 136–141. Boston, Massachusetts, USA (2006)
37. Stenz, A.: The focussed D* algorithm for real-time replanning. In: Proceedings of the International Joint Conference on Artificial Intelligence (IJCAI), pp. 1652–1659 (1995)
38. Sturtevant, N.: Memory-efficient abstractions for pathfinding. In: Proceedings of the third conference on Artificial Intelligence and Interactive Digital Entertainment, pp. 31–36. Stanford, California (2007)
39. Sturtevant, N., Buro, M.: Partial pathfinding using map abstraction and refinement. In: Proceedings of the National Conference on Artificial Intelligence (AAAI), pp. 1392–1397. Pittsburgh, Pennsylvania (2005)
40. Sturtevant, N.R., Bulitko, V., Björnsson, Y.: On learning in agent-centered search. In: Proceedings of the 9th International Conference on Autonomous Agents and Multiagent Systems (AAMAS) (2010)

Embedding Information into Game Worlds to Improve Interactive Intelligence

G. Michael Youngblood, Frederick W.P. Heckel, D. Hunter Hale, and Priyesh N. Dixit

Abstract Current game worlds are visually rich but information poor – particularly poor from the artificial intelligence (AI) point of view. Where the player sees a rich visual representation of 3D objects, internally these are just very sparsely described collections of points in space. Tools for advanced world creation, character modeling, animation, and advancements in computer graphics have brought us into the age of near photo-realistic interaction; however, these interactions are still very limited in comparison to the real world, and the information is presented overwhelmingly for the player, packaged for the Graphics Processing Unit (GPU) with little reflection or structure suitable for use by AI systems. This problem of a lack of rich information suitable for consumption by the game AI often limits the true potential for deeper levels of interaction that are becoming more in-demand by game players. This chapter presents a number of tools and techniques, which are being used to improve the embedded information contained in immersive game worlds. Symbolic annotation of the environmental elements, advanced spatial decomposition, calculating the information value of the surfaces in an interactive environment, and visual analysis form the core tools and information generators of our Common Games Understanding and Learning (CGUL) Toolkit. Using these tools to incorporate information into the game design and development process can help create information-rich interactive worlds. AI developers can work with these environmental information elements to improve non-player character (NPC) interactions both with the player and the environment, enhancing interaction, and leading to new possibilities such as meaningful in-game learning and character portability. Case studies from two different projects using these techniques provide some additional insight and reference as to how these techniques have been incorporated into current game AI and research.

G.M. Youngblood (✉)
The University of North Carolina at Charlotte, 9201 University City Blvd., Charlotte, NC 28223, USA
e-mail: youngbld@uncc.edu

P.A. González-Calero and M.A. Gómez-Martín (eds.), *Artificial Intelligence for Computer Games*, DOI 10.1007/978-1-4419-8188-2_2,

1 Worlds of Visual Richness and Information Poverty

Creating virtual worlds for games, simulations, interactions, observations, or any of the varied applications of these constructs is becoming a common task for the masses. This type of creation activity was once the purview of skilled modelers and trained digital artisans, but since the turn of the twenty-first century the barrier to entry has been significantly lowered. Creating virtual worlds is more accessible now than in times past, but even more amazing is that the graphical fidelity has increased while making this task easier. It has been the creation of tools that has facilitated this change—tools with different specialties for each part of the virtual world creation pipeline.

The first step of the pipeline is modeling, which is the creation of three-dimensional (3D) virtual objects in a tool that allows the combination and contortion of primitive shapes into complex models. Models defined by dozens to millions of vertices can be created using modeling tools such as Autodesk's Maya (`www.autodesk.com/maya`), 3D StudioMax (`www.autodesk.com/3dsmax`), or Softimage (`www.autodesk.com/softimage`); NewTek's Lightwave 3D (`www.newtek.com/lightwave`); the open source Blender (`www.blender.org`), or the freely available Google SketchUp (`sketchup.google.com`). These tools provide a means for modeling, animating (key framing and interpolating motion between frames usually based on articulated skeletons associated within the models), and rendering (the final target visual representation). In this chapter, we will talk about the modeling and animation aspects, but will assume that rendering will be done in the target interactive environment (i.e., the video game or simulation). We focus our terminology on polygonal modeling, but other types of modeling (e.g., NURBS) have similar properties and the reader should apply the same ideas regardless of the underlying modeling paradigm.

Modeling includes more elements than just the arrangements of vertices in space to create the desired geometry of the object being modeled. It also includes more than just static geometry (e.g., buildings and terrain) as many elements are specifically created to be dynamic (e.g., doors in buildings, characters, vehicles, and the ever-popular gun turrets). Models are textured with colors or imagery (i.e., 2D art) which is created separately and then *projected* upon the surface of the model. The base texture, referred to as the *texture*, is defined as the 2D pixel array containing the colors to be applied to the target model by a projection (e.g., spherical, cylindrical, or UV coordinates that map to specific vertices of the geometry). Other types of texture *maps* are also applied using the same projection. These include normal maps (a.k.a., bump maps) that define the surface height details by defining the normals, or slopes, of the surface allowing for light and shadow to be rendered appropriately giving depth to the base texture (height maps can also be used), specular maps that define the shininess of the base texture at specific locations, and diffuse maps that define how light reflects both in color and intensity. Other types of maps exist such as hit maps used to determine locations of damage, maps to aid in rendering infrared images showing the heat signatures of modeled objects, and other maps useful for representing various phenomena on the surface of the model. Textures and texture maps are typically created with 2D image tools such as Adobe Photoshop (www.adobe.com/photoshop) and/or The Gimp (www.gimp.org), but are increasingly being made through procedural tools such as Allegorithmic's MaPZone (www.mapzoneeditor.com). The model is usually saved as a mesh containing the vertex and edge information along with the texture and texture map file names and their projections onto the model. Thus, a model is saved in the model format along with the set of textures applied to that model (multiple files).

Dynamic models move and usually have one or more articulated parts. The modeling tools often allow the user to specify the skeletal structures of objects (any objects, not just simulated organic creatures) and define how the joints operate. Using a timeline, the user can create key framed poses to animate along the defined degrees of freedom. Forward kinematics is used to calculate the positions of the object between the key frames. Location and orientation in space can also be set in addition to the skeletal pose. Another type of texture map is used to define the weights of the mesh shell around the model skeleton specifying how it deforms with movement. The *animator* usually names each animation sequence as a set of frame numbers, which are typically saved along with the model in the model file. This process can also be performed with mesh vertex movement over time (i.e., vertex animation instead of skeletal animation).

The modeling tool can also make *scenes* that consist of several models arranged in space. However, these scenes are typically used for rendering and not for virtual world creation for games or simulation (although they could be used for this task) using the tool's renderer for movie or image generation. A game engine-specific *level editor* is typically used to assemble components and models into a world for interaction. Tools such as Ambiera's Irredit (www.ambiera.com/irredit) for the Irrlicht 3D Engine (irrlicht.sourceforge.net); QuArK (quark.planetquake.gamespy.com) for about 45 popular games using the Source (source.valvesoftware.com), Id Tech (www.idsoftware.com/business/technology), and other engines, as well as other editors often packaged with the PC version of a game that are used to create the immersive virtual worlds generated by those engines. Beyond just the arrangement of models in space, these level editors usually allow for some geometry creation as well as setting up light, dynamic elements (e.g., doors), and spatial logic (e.g., an area trigger that will open a gate when the player enters). Engine-focused level editors consume models from a modeler and allow the designer to assemble the interactive virtual world. The accessibility of these tools has inspired many individuals to create content and even whole games. The Mod(ification) Community consists of thousands of people worldwide who as a hobby make new and interesting virtual worlds for others to experience. Even online in persistent worlds such as Second Life (secondlife.com) user-generated content including buildings, clothing, and just about anything imaginable is available and is being created by the user, not the developer.

The problem with all of these tools in the games and simulation content creation pipeline spanning texture, texture map, model, animation, and level creation is the incorporation of information elements with each of the products. Creating virtual worlds for people has been the primary task for which these tools have been created, and they do a tremendous job at providing visually rich worlds for humans. However, humans are not the only elements interacting in these worlds as more and more intelligent characters are acting as friends, foes, proxies, and other integral roles within these virtual worlds. In games, particularly single player games, the non-player characters (NPCs) are the major actors in creating the drama and interactions to make it a game. Even in multi-player games, NPCs (also known as bots) round out teams and increase the immersion and action in the first- and third-person player genres. In the strategy genres (i.e., real-time strategy, turn-based strategy), the NPCs take and perform actions often at a high level of description from the players – making everything happen in the game. The core elements of interactivity in these environments are currently very controlled and explicitly designed and implemented, but the demand for increased interactivity and deeper immersion is on the rise. One mechanism to increase the fidelity of interaction in virtual worlds is to improve the intelligence and behavior spectrum for the NPCs [i.e., improve the artificial intelligence (AI) driving the game characters]. The first place to start is to give them more information about the world in which they are interacting. We must move from our current information-poor worlds to information-rich worlds, and since the user is becoming a key contributor, we must provide them with the ability to improve the information of their contributions as well.

The current problem of perception is illustrated in Fig. 1. All of the models, textures, and the arrangement rendered by the engine display objects recognizable to a human, but to the machine they appear as a set of points in space with some connectivity and set relationships. The machine cannot easily determine what it is perceiving is a house or a tree or a dog or a spaceship or anything really. This is the same problem that machine vision researchers are trying to solve – recognizing objects from images. However, in virtual worlds we do not have to solve this problem, all we need to do is to incorporate some knowledge engineering tasks into the content creation pipeline, namely, we need to have the creator add information elements that are attached and persist with the models, textures, and worlds.

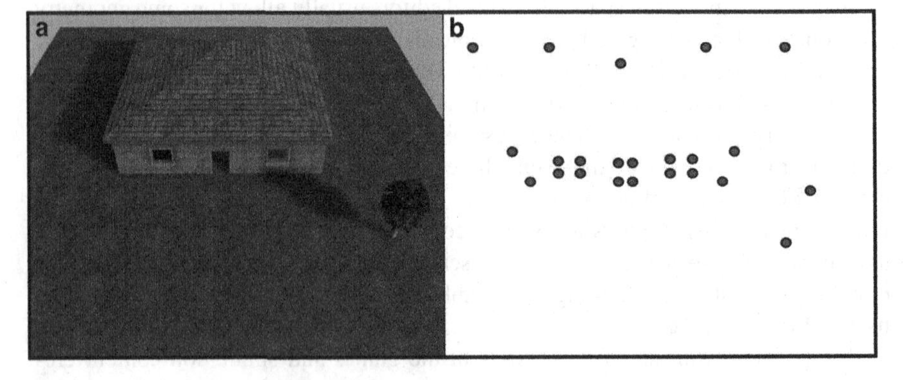

Fig. 1 In a virtual world, the human easily perceives the items in (**a**) as a house and a tree, but to the machine the world appears more like (**b**), as a collection of points in space

For example, when the modeler makes an oak tree they need to annotate it as an *oak tree*, or perhaps more specifically a *post oak tree from Texas*. The texture and texture maps could also contain this type of symbolic information, which could be consumed by intelligent entities in the virtual world.

Our focus is on techniques and tools that can be used to add information to elements of a virtual world targeted for games and simulations, but this work can certainly be more broadly impacting. In the sections ahead, we present tools and techniques for embedding information into virtual world objects (both static and dynamic) and their surfaces that lead to improved knowledge for consuming and interacting with elements of these environments with the goal that artists from professionals to hobbyists can incorporate these into their virtual world creation pipeline.

1.1 The Need for an Information-Rich World

Our own research work is in the field of game artificial intelligence (game AI) where we are in pursuit of human-level intelligence at both the character level and the system level. There is much work to do in this area, but we are strong supporters of the idea that interactive computer games are an opportunity area for exploring techniques and theories leading to human-level AI [14, 16]. One of the key problems in our current information-poor virtual worlds is that the playing field is clearly not level for AI characters (NPCs), especially when competing with humans in the same space. Hence, in practice game AI is often implemented with a level of cheating such as using perfect information (i.e., complete world knowledge) and unseen perception-aiding techniques (e.g., breadcrumb trails to follow the motion of characters). The first step is evening perception, which will allow artificial characters similar sensory perception of the world that the human players have. The human brain is a tremendously good pattern matcher to known objects; so most players are entering into these virtual worlds with 2–80+ years of real world experience. This is a little hard to compete with from the game AI character perspective, especially since the field of AI is just a little over 50 years old itself. However, what is really needed is to at least present similar information to both the human and non-human players, in this case, presenting object information and spatial relationships in a more readily processed manner to the NPCs – the same information the human gets with their eyes and from object recognition with the knowledge they acquired over time. We can then leverage the intelligence of the game AI designer/developer to encode proper responses to perception.

An interesting trend over the past few years has been the emergence of 3D object warehouses that offer downloadable content for free (`sketchup.google.com/3dwarehouse`) or purchase (`turbosquid.com`). Sites such as these make tens of thousands of models available for use in virtual worlds, rendered scenes, and any use imaginable. The vast majority of the models made are created by freelance, hobbyist, and even amateur modelers (many are students learning this art form).

It is commonplace to just buy off-the-shelf models to use in projects; however, this continues to propagate and even worsen the problem of information poverty. Given that embedding information into these creations is currently not a standard practice or even a step thought about in the process, there is now a massive, growing number of potential virtual world elements that have no information annotations. Furthermore, once the models are sold as a product, contact with the original creator diminishes and so does the full meaning and inspiration information elements associated with the creation and the creation process.

From a character AI development standpoint, information-rich elements assembled as the worlds are created can flow their specific information into tools for behavior development and be presented in-game to the AI character as perceptions of those objects. This is the key to helping create better AI-driven characters, providing a flow of meaningful information to better perceive the world objects, determining how to apply actions from the repertoire of available actions of the AI characters, and gaining some understanding from the elements in their world. We all make better decisions when better informed.

Acquiring knowledge over time and being able to apply previous knowledge from old problems toward solving new problems is the process of learning. Having more information-rich worlds will also allow NPCs the ability to apply machine learning techniques over observed sample data from observing other players (human and artificial) as well as itself. Being able to expand the knowledge of characters and allowing them to adapt over time is an exciting potential for interactive characters, which could increase the immersive experiences of virtual worlds.

The sections moving forward are being written as a way to educate the current and new masses of modelers and tool builders with regard to elements important for game AI, providing insight through a particular lens for fellow game AI practitioners, and establishing a way to move toward common approaches of encapsulating information elements for interactive (game) AI in virtual worlds. This is also a blueprint for the interactive AI researcher and developer to better understand and help guide the incorporation and usage of information-tagged elements. Beyond just information tagging, we will also discuss ways in which the information could be organized and how additional information can be processed by leveraging information-tagged elements.

1.2 Overview

This chapter presents a cohesive and integrated view of our key approaches, ideas, and work in adding, generating, and using information in interactive virtual worlds. We will begin with basic information tagging to objects, present a method for organizing information, discuss the generation of knowledge from processing world information, and then provide some insight into research projects that have incorporated these techniques and ideas. Throughout the discussion key principles will be highlighted to allow for easy incorporation into one's own workflow.

2 Embedding Information

The fortunate byproduct of the the adoption of standards such as XML is that ASCII-based, readable formats are in vogue.

This section will present a number of ideas with references to techniques and tools that can be embedded into the comments or extensible sections of many file formats. Keeping information as close to the original element as possible is preferred, but care should be taken to make sure that the added information is preserved when editing the source.

Key Principle: Prefer embedding information as comments or extensions in the source files over creating separate information files.

2.1 Information at the Object Level

The lowest level of embedding information should be at the object level. Each model should be tagged with information that describes what it is (i.e., nouns) and the properties of the object (i.e., adjectives). As each object also provides some set of actionable properties (i.e., a verb can be applied to it), an affordance that defines that property between the object and an actor (e.g., AI character) should also be added. Affordances are often described merely as the action(s) that can be applied (e.g., an object an actor can sit upon is described as having the affordance "sit"). The term *affordance* was introduced by Gibson to describe all "action possibilities" available in the environment [6]. For example, if there is an unoccupied chair sitting upright in the world, there is also an opportunity for sitting in the chair. In game AI, this can be used to sidestep the symbol grounding problem, at least in virtual environments; rather than define the quality of *chairness*, we merely require that anything that can be used as a chair should have a "sit" affordance defined.

Affordances can be directly added during creation or level design to objects in games to provide information about what sorts of actions can be taken using the objects to which they are attached. Affordances may also exist separate from objects in the environment. If a character possesses a hang glider, for example, it will state that it can be used as a tool to enable flight, but that it requires a launch point. A launch point affordance could then be placed at spot on a cliff edge. By the level designer placing this information in the world, the character AI needs no additional knowledge about launch points, or even the difference between a hang glider or a paraglider. Affordances have been successfully used in several games and simulations, including the popular Sims series from Electronic Arts [4].

Formally, we can define the information to be stored in each object model as the 3-tuple (ν, α, Δ) where ν is the set of all describing nouns (descriptions), α is the set of all describing adjectives (attributes), and Δ (change) is the set of all affordances. Note that applying an action to an object may change the state of an object (e.g., change its location or potentially damage it). An embedded synonyms list

Fig. 2 Objects tagged with Cyc constants (identifiable by the hash-dollar prefix) in a virtual world. Note how more perceivable the world is from an information view. Note that not all objects are labeled in this figure, but from what is, the character AI could start to reason about this world. Canine AI could head toward the tree, character AI could head toward the door and enter the house, and so forth – a world of many possibilities

could also be kept for each description to facilitate better understanding and faster searching for matching concepts or actions, or a lookup using a tool such as WordNet (`wordnet.princeton.edu`) could be used. Foreign language equivalents could also be embedded or looked up using commonly available translators.

The descriptive tuple elements can be filled out using the standard words and grammar of the person providing the tags (preferably the creator of the object) or using descriptive elements from a common sense knowledge base and a relational ontology such as those provided by ConceptNet (`csc.media.mit.edu/conceptnet`) or Research Cyc (`research.cyc.com`). An example of Cyc constants (nouns) tagged to object models in a virtual world can be seen in Fig. 2.

Some objects are more complex to model than others, and it is commonplace to model complex objects as a series of connected parts – particularly objects such as articulated parts (e.g., the turret of a tank is typically modeled separately from the body). Each part should contain its own set of annotations as well as the whole object containing a traceable reference to the information of the constituent components. In many game/simulation engines, the model component names can be retrieved and associated with the information tagging elements, but creators should be careful to ensure that the names match both in the information tags and the model part names.

Key Principle: Embed description, attribute, and affordance information within each object part and the whole object.

Another factor in creating virtual environments and objects is the way in which the world elements are built, particularly structures that can contain other objects. Modeling tools typically allow the creation of objects such as extruded rectangles and then allow the user to remove a portion of it as illustrated in Fig. 3a, b. If the model was just for visualization, this is perfectly acceptable; however, when the model is used as an interactive element or an element that will be perceived by game AI, it can cause problems. If there was a window in the void, it would appear to be placed in the middle of a wall with a section cutout, and this causes an

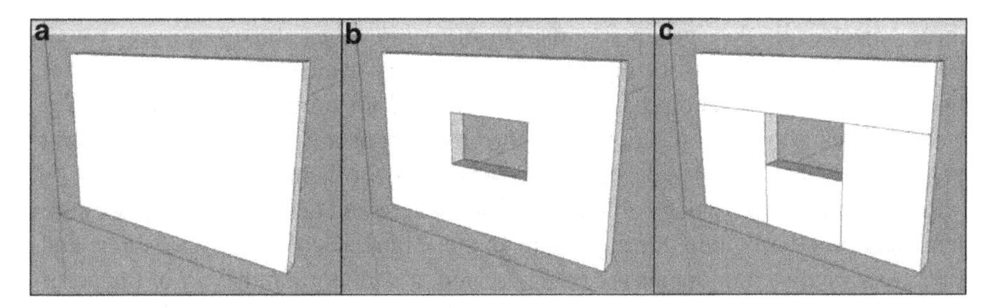

Fig. 3 Using a modern modeling tool (Google SketchUp in this case), a user can easily create a 3D object such as the wall in (**a**), but when creating a window as in (**b**) the tools can cause problems from an information standpoint. Panel (**b**) shows an extruded rectangle with a cutout missing extruded rectangle making the hole, which can be difficult to reason about spatially especially if a window occupied the missing cutout. From a game AI and information tagging perspective, it would be better to model each section of the wall independently with regard to potential physics engine forces as shown in panel (**c**)

unnecessarily complex spatial arrangement that may have to be reasoned over by an NPC. It also causes problems for information tagging as there is a void area to negate the information and then replace with potentially another object's information (e.g., a window placed in the void). A more simple approach would be to model the wall and void as shown in Fig. 3c, which allows each segment to stand on its own with its own properties without any complex spatial arrangement or potential overlap in void/occupied space from multiple models if a window was installed. This arrangement also shows a structure that could perform more consistently to a real structure under the influence of forces from a physics engine – something that should also be taken into consideration.

Key Principle: Keep world geometry simple and avoid the use of object cutouts.

Remember that there are two types of objects in interactive worlds: static elements that do not change state and dynamic elements that change state in many degrees of freedom. The principles described to this point apply to both types of models, but dynamic objects should also contain information pertaining to their state changes and degrees of freedom to provide information to intelligent entities. Human players can easily recognize a door in a building or house and have an intuitive feel for the actions and behaviors it will have under dynamic conditions; however, NPCs have no idea – How far does the door open? in what direction? and so forth. Dynamic information can be represented by a set of variables, their ranges, and actions that can be applied (affordances). Since physics engines are used in modern games and simulations, all world objects should have a defined mass (and distribution if not uniform), weight, and center of gravity. This information can assist in reasoning about these objects and how they will behave when actions are applied to them.

When creating world objects, simplifying bounding geometry should be provided to the game/simulation engine for use in collision detection and other calculations. Default settings in most modeling tools will provide bounding boxes over the whole object or the key grouped parts, but oftentimes these boxes are very rough simplifying boundaries. Ill-fitting bounding geometry can cause problems with not allowing objects close enough, creating odd force fields around world elements, or in other cases allowing objects to pass through supposedly solid objects. Content generators need to include good bounding geometry for their creations.

Reuse from prior projects, off-the-shelf purchases, and using open work are core elements in modern game development. As these libraries of models, textures, and other game-ready components are being created, it is important that *all* of the pertinent information be added to these objects. Embedding the information described in this chapter can greatly facilitate use of created objects into games and simulations because they lend themselves to supporting game AI and even game physics. Modelers at all levels need to think beyond just the visual aspects and be sure to provide *information complete* objects. This notion of *information completeness* is an important concept and may be a deciding factor in whether or not a modeled object can be used in a game or simulation, especially as game AI increases in complexity – in fact, making more advanced game AI depends on this in many cases.

In the next sections, we will discuss a number of techniques that have tools associated with them that were developed from researched techniques discovered by the University of North Carolina at Charlotte's Game Intelligence Group (`gameintelligencegroup.org`). If the reader is interested in any of these tools or learning more about these techniques, please refer to our website or contact us directly. Many of these tools, while not available to the public, are shared with fellow academics and can be licensed to industry.

2.2 *Creating Places*

Creating virtual worlds or levels by assembling and placing models in the desired configuration is only the beginning of the work needed to make a viable interactive environment. Even with *information complete* models, which include proper bounding boxes for collision detection and resolution, we only have an environment that player-controlled characters can explore. Character AI could carefully stumble around exploring and creating a map in its memory, but this would take time and be rather inefficient. What is needed is a way to represent and understand the free space between objects in the virtual world to support spatial planning and pathfinding.

There are a number of representations and methods for decomposing the navigable free space that exists between and inside virtual world objects. These range from simple graph-connected, manually placed way points to advanced automatic decomposition techniques [10]. While way points have their uses, automated spatial decomposition methods can provide navigation meshes (navmeshes) that provide higher coverage of the navigable spaces and a multi-modal method allowing for reasoning at both a local detailed level and a topological view. Many techniques based on Graphics Processing Unit (GPU)-like decompositions, vertex connectivity, and Delaunay triangulation decompose the environments into lots of low order shapes (polygons in 2D and polyhedrons in 3D). Low order shapes tend to have many triangles or pyramids converging at common vertices that make localization near those points difficult due to character footprint.

Human navigation is not based on thinking about space as moving from one precise point to another and does not have convergence difficulties in the corners. We tend to view the world as a series of connected places. This place-centric notion can be very helpful when creating navigating character AI because we can match the navmesh representation to our own intuitive view of the world – one in which we can more naturally understand and more intuitively specify directions to the character AI. Higher order decompositions such as those we have developed, which are based on polygon (2D) and polyhedron (3D) seeding and growth [7–10], can provide complete world coverage. Decomposing all open spaces with regions that touch each other allows us to create a topological map of the region connectivity, while also providing a region breakdown that more closely represents a series of connected places. Figure 4 illustrates some of the basic methods of decomposing navigable space into a navmesh. We advocate using our Planar Adaptive Space-Filling Volumes (PASFV) algorithm (2D) and Volumetric Adaptive Space-Filling Volumes (VASFV) algorithm (3D) to create places and the associated navmeshes within a created world. Navmeshes allow for more efficient path planning using search over the topological graph and easier pathfinding by reducing local navigation reasoning to smaller, more manageable places. The decomposition of space can also be leveraged to reduce reasoning complexity by compartmentalizing information into the places where the objects exist. This notion of *information compartmentalization* is useful in that it reduces the number of objects and information elements an AI entity has to reason over to those in its immediate place and possibly the adjacent places.

Information complete models can aid in decomposition methods by biasing the algorithms toward common elements such as areas covered by grass, sidewalk, or road. In this manner, the areas covered by like objects and materials can be cohesively decomposed making it easier for character AI to "walk on the sidewalk" or to "play in the grass." This generated information now added to the world can also be organized hierarchically to facilitate improved efficiency in spatial planning or

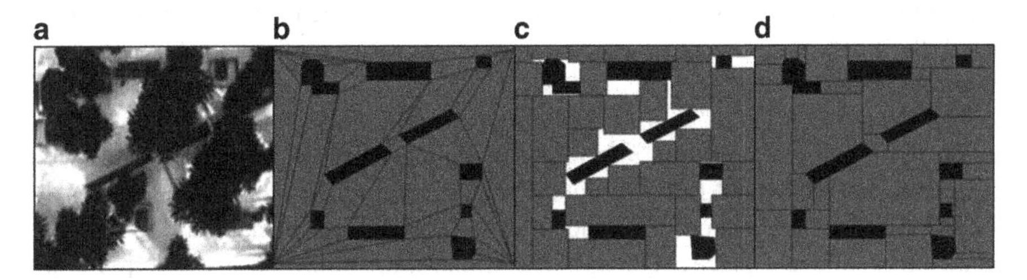

Fig. 4 (**a**) A game level from a Quake 3 Mod (Twin Lakes from Urban Terror by Silicon Ice Development on id Software's idTech 3 Engine). (**b**) This level decomposed with the Hertel–Mehlhorn algorithm [15], a triangle-based vertex-connecting technique. (**c**) Same level decomposed with the Adaptive Space-Filling Volumes algorithm [19]. (**d**) Same level decomposed with the Planar Adaptive Space-Filling Volumes (PASFV) algorithm [9]

other relevant tasks. In addition, it makes it easier to pre-compute and store actions within those places as is a common technique in game AI for increasing speed and efficiency.

Key Principle: Leverage knowledge generation techniques such as spatial decomposition to help provide deeper world knowledge in an intuitive, useful, and efficient organization.

2.3 Information Services

All of the information we have embedded, then generated, and finally organized is typically collected and then presented to requesting components of the game engine. The common method for delivery of this type of service is through a code library that provides the desired functionality in linked code, since games are typically created with compiled languages. However, it could also be provided through interprocess communication (IPC) mechanisms with defined protocols – we have provided services using shared memory in a number of game environments as well as with code libraries. We have found it helpful to write processing elements that collect and organize all of the information from the source materials into two files, one for static information that never changes and one for dynamic information that has to work with the game engine more closely. The part of our Common Games Understanding and Learning (CGUL) Toolkit that provides the static information is called Static Spatial Perception Service (SSPS), and the part that provides the dynamic information is called Dynamic Information Augmentation Service (DIAS).

SSPS (pronounced "sips") at its core provides access to the navigation mesh and all compartmentalized object information. The mesh itself is represented in two ways: as a collection of geometric entities and also as a directed graph. The geometric representation is useful for local place navigation and storage of compartmentalized information, while the graph includes search functionality (A* search for CGUL SSPS) to allow global navigation. When SSPS loads, it reads the navigation mesh and all object information from an XML file. It then constructs a default graph to plan through a defined size character bounding box. When a character is instantiated, the SSPS service can generate a new graph based on the actual size of the character bounding box.

The main SSPS query is *findPoint(location, hint)*. This query searches for *location* in the navigation mesh, returning the unique ID of the place in which it is contained, or `UNDEFINED` if it is not contained in any place. The *hint* provides a starting point for the search; if provided, this place will be checked first, followed by all of its adjacent places. The search continues, breadth-first, until the point is found or all places have been checked.

Search-based (A*) path planning is provided by the *findPath(start, end, start_hint, end_hint)* query. This query returns a *list* of place IDs. If no path is found, the returned list is empty; otherwise, there is a path through the navigation mesh from the starting point to the end point that passes through each of the places in order. Character AI can use this to traverse the maps moving from place to place. Navigation between places involves moving from *gateway* to *gateway* within the place by local navigation. *Gateways* define the connection planes between two places. Local navigation can use simple centroid movement with obstacle avoidance, probabilistic roadmaps, occupancy grid navigation, or any reasonable method. Due to the smaller size of a place in comparison to the world, local navigation is usually fast even using heavy-weight techniques.

Several other SSPS queries exist to support other needs. For example, *generateRandomValidPoint(region, ground)* will generate a random point, guaranteed to be in a free space (navigable) region. The region reference is filled with the ID of the region in which the new point is contained. The ground parameter is a boolean value specifying whether the point should be on the ground; if it is false, the generated point will have a z (height) value between the region's z_{min} and z_{max} values. If a point is believed to be within a positive space (non-navigable) region, the *findPointInObject(location)* query will search all positive space regions for the given location.

The primary computation performed by SSPS is the point in polyhedron calculation, which tends to be very fast once localized, given that the character movement is not too rapid or that they cannot teleport across the world at random. As the character is localized, information about all objects (hopefully information complete objects) in the current place is provided and other information is available by query (e.g., information about adjacent places). Thus, SSPS provides a constant, detailed perception of all static objects in the virtual world to interested entities in the game/simulation engine.

DIAS is a similar service to SSPS, but it only provides additional information about dynamic objects. The core game/simulation object usually maintains all dynamic information since it controls their behavior and state in accordance with their design and the engine design paradigm. However, to convey model and other external information (e.g., affordances), DIAS serves to augment the internally tracked dynamic information for the object. It can also be used to help interpret specific engine implementation states to a more common state model to present to the game AI.

Key Principle: Utilize services to provide the *information complete* data to key game/simulation engine components and facilitate the use of more advanced AI techniques to improve both character and system intelligence.

2.4 Virtual World Dynamics

The environments of modern games and simulations are ever changing. As both players and characters interact with each other and the objects and environments of the virtual world, context changes often govern new modalities of play and interaction. As more *information complete* elements comprise these virtual worlds, the dynamic nature and potential only increases. We discuss two primary notions of handling virtual world dynamics to support game AI, namely *probabilistic affordances* and *influence points*.

2.4.1 Probabilistic Affordances

The basic concept of affordances is useful in itself as we previously discussed, but it does not address differences in characters, or how world state may affect affordances. We approach these issues with the idea of *probabilistic* affordances. *Probabilistic affordances* may present different actions depending on the current context of the interaction.

Our approach is to create affordances that use information about current world state and the character AI to determine the currently available set of actions. A probabilistic affordance is a state machine with an input set composed of environment and character attributes, an output set of actions, and an output set of expected outcomes. Environment attributes include information about the local place: what characters are present, their relationships, and the state of any doors, objects, or other devices in the place. Character health, inventory, action model, and history of interaction with related affordances are part of the character attributes. The character action model in this case refers to the actions that the character is capable of performing. In addition, character attributes may include physical or personality characteristics. Physical characteristics may be role-playing game (RPG) style attributes such as strength, dexterity, and wisdom. Personality characteristics define how characters will react in social settings and may be a character's alignment in an RPG or a set of Hofstede parameters in a cultural simulation.

The actions output by the affordance are from the character's action model and include how the object or world location the affordance is tied to should be used. For example, the action output for pulling a lever would include three pieces of information: the *pull* action, the lever object as the *target* for the pull action, and finally the location the character should be standing in to pull the lever. For a sword on the floor, the character would see the *pick up* and *attack* actions, with the sword as the *target* for pick up, and the sword as the *tool* for attack. The pick up action would include the location of the sword, but the attack action would not provide a location.

Finally, the expected outcome distribution provides the character a notion of the expected change in world state after the action has been executed. This is a probability distribution to allow the specification of outcomes that are uncertain. In the previous examples, the lever pull action would have two probabilities: the linked door could open with a 90% probability, the rusty chain that raises the door could break with a 7% probability, or the lever itself could break with a 3% probability. For the sword, the pick up action would place the sword into the character's inventory with a 95% probability, or the character could trip and fall with a 5% probability. Until the character picks up the sword, the attack action will have no probability of success, but once the character has the sword, it will provide probabilities based on the characteristics of the sword.

Note that the affordance input requires a history of interaction with related affordances. This allows additional interesting scenarios. Imagine that our hero has just entered a dungeon room, and in this room there are four levers, three doors, and a trap door leading to the pit of infinite doom. The affordances on the levers have outcomes of opening the door to the exit, the treasure, or a peckish dire badger. In addition, they have a probability of triggering the trap door to the pit of infinite doom. As our hero pulls the levers and observes the results, the probabilities should adjust to reflect his experiences so far. Including the history of how the hero has interacted with each of the levers with the query allows the affordance probabilities to adjust as he tries each lever. If the first lever opens the exit, the probabilities are now 33% finding the treasure, 33% encountering the monster, and 33% certain death. Deciding that this is an acceptable risk, the hero chooses to open one more door. His second attempt releases the dangerously hungry dire badger. After defeating the monster, the hero can decide that his threshold for danger has been surpassed, and he is not willing to risk the 50% chance of falling in the pit for the chance to retrieve the treasure. If our hero fails and falls into the pit of infinite doom, the history of his interactions with this set of affordances is carried with him to his next life. So long as the history of interactions is maintained, no other changes need to be made to the hero's controller to allow him to make the correct choices when he next enters this room. In effect, learning occurs, but without the need to explicitly model a learning process for the character. This updation also applies to interactions with players and how they change the environment.

Some actions are only possible with the assistance of others. Affordances as discussed so far only support a single character; one character queries the affordance for the details of the actions that are possible, and then can attempt them alone. We extend the notion of probabilistic affordances to provide multiple *slots*. A slot is an action that one character can take to activate the affordance, and multiple slots allow multiple characters to work together to achieve some action not possible with a single character. Each slot is either required or optional. Required slots must be filled for the action to be successful; optional slots may enable it to complete faster or provide a better chance of success for results that are uncertain.

To return to the levers example, a multiple character affordance might be required to pull a particularly rusty lever. This lever has three action slots, two of which are required and the third is optional. One character does not have the strength to pull this lever, but two can. However, if only two characters attempt to pull the lever, it only has a 60% chance of working. A third character can join and improve the chance of the lever pull opening the door. Thus, this can help facilitate collaboration between not only NPCs but also players and NPCs.

Probabilistic affordances start with the artist annotated affordances for an object and are then extended by the game/level designer into the probabilistic affordance framework, which takes into account the multiple contexts and requirements for using and interacting with virtual world objects. This also puts a significant portion of the prerequisite information for intelligent use of an object with the object in the place it resides. This reduces the burden for remembering/storing all world interactions with the character and distributes it throughout the virtual world to be retrieved at the proper time under the proper context.

Key Principle: Probabilistic affordances should be added in the annotation phase of modeling and/or level design, but can update through interaction providing the ability for game AI to change with virtual world dynamics.

2.4.2 Influence Points

Influence points provide a technique to add information to the game world at run-time so that characters can take advantage of new information that was not available at the time the world was created. The primary use of influence points in our research has been to allow the addition of tactical information to the environment by enabling characters to use the world as a blackboard for communicating observed conditions to teammates. Influence points are similar to the idea of influence maps, but take advantage of navigation mesh decompositions – which we highly encourage as a spatial navigation structure as previously discussed. Individual influences are set in a specific place instead of using a high-resolution grid map. Because a place includes an area that would be covered by a much larger number of map cells, this allows much greater efficiency than influence maps. The influence of an entity on a map may not be limited to a single region – in these cases, additional child influence points are added to neighboring regions. To demonstrate the idea, we present an example of how characters can use influence points to further enrich the world with information.

A character is attempting to make it to a goal, the gold mine, to fetch resources as shown in Fig. 5. The character starts at its home base and has a choice between two paths. The longer path is defended by a friendly turret. The other path is short, but unknown. The character travels the short path and is ambushed by enemy forces. After retreating from the battle, it updates the navigation mesh by adding an enemy influence to the area where it was attacked.

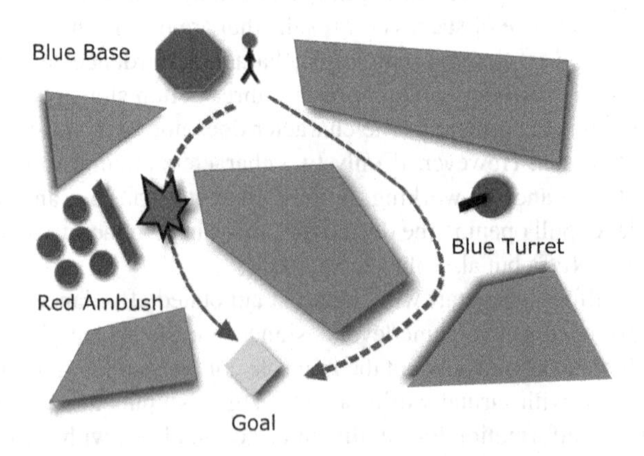

Fig. 5 Example world in which influence mapping can provide useful tactical information

Now the path passes through an area where the enemy is powerful. The longer path, on the other hand, passes through an area that is well protected by a defensive turret. After running its pathfinding algorithm a second time, the character finds that the least expensive path (incorporating chance of injury as a cost) is the longer path that passes the defensive turret. The character now follows the longer path and makes it to the gold mine successfully, returning valuable resources to home base. Other characters can now use this information as well, as it is maintained in the world representation rather than in the initial character's internal memory. The point strength of the influence decays over time to represent uncertainty about changes in the world, but it can be updated after creation.

In this example, if a grid-based technique is used, the character must spend a nontrivial amount of time inserting influence into the map, and potentially updating the large number of cells affected by the enemy and ally forces. If a navigation mesh is used, however, the insertion and update steps could be far less expensive; insertion is $O(n)$ in the number of regions in the navigation mesh, and updates to influence values can be made in constant time for decreasing values [or $O(n)$ for increasing values] [12].

While our example shows the influence points used primarily as a tactical tool, they can also be used to provide other types of information. Characters can label the world with information about events that have occurred or mark a region as a target for a particular activity. For example, if a character makes a mess in a virtual kitchen, it could place a "dirty" influence in the room. This would mark the kitchen as a target for another character to clean, or get into an argument with its virtual roommate/spouse about housekeeping responsibilities.

Key Principle: Dynamically embedding information into the environment at run-time can greatly improve character intelligence by adjusting to changes within the game/simulation.

2.5 *Information from Interaction Observation*

Running analytics on every aspect of a game including all player and NPC decisions and actions, as well as tracking spatial movements to understand level utilization and overall performance, is a popular focus in today's game industry – especially in online games where keeping interest high keeps paid subscribers engaged. The key to this is logging. Log everything useful that does not affect game play by slowing down the system, and with user permission collect these data to improve game play and the user experience.

Key Principle: Log and collect as much relevant data from game play and interaction as possible without impacting game performance.

Collecting data is just the first step, what is interesting is what additional information can be learned from that data. In many cases, there is so much data that it can

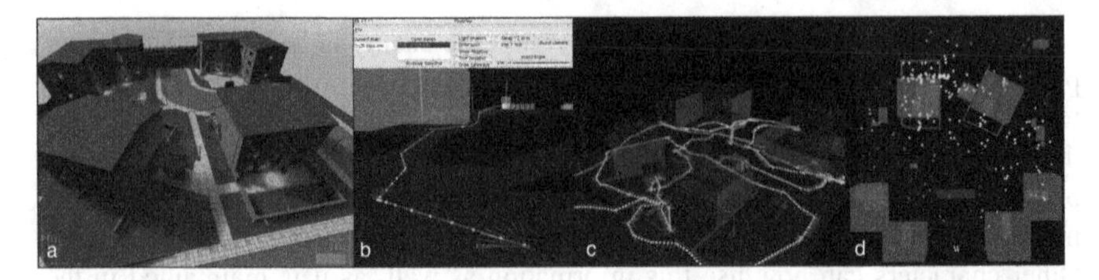

Fig. 6 Example analytics derived from logged data collected from the (**a**) Urban Combat Testbed and visualized using (**b**) PlayerViz showing the player traces and interactions over time for (**c**) an entire play session. This can be further analyzed through visual data mining, developing math models to automatically create (**d**) heat-maps showing the commonality of behaviors in spatial locations across a number of players/characters

become unwieldy to work with and gain knowledge. Visualizing the data and using an interactive visualization process to manipulate, isolate, enhance, and frame the data, such as is provided by our PlayerViz tool (shown in Fig. 6b), can be useful for performing visual data mining to find patterns in the data. Math models can be built around discovered regular patterns and used to generate other visualizations such as heat-maps (shown in Fig. 6d) of phenomena occurrence in the virtual world [18,20]. Such analysis can be useful for understanding interaction problems, exploits, use of environmental elements, and problems with level design. The key unit of analysis we use is the *player trace* (shown in Fig. 6c), which captures player/character spatial motion colored over the passage of time, gaze direction, and any interactions (e.g., pushing a button, firing a weapon).

Player traces that capture the view frustum of the player can also be used in conjunction with world geometry to track what surfaces the players viewed. Over a number of players, we can use a texture painting technique to radially and distance weigh observation from the virtual fovea point to the incident world geometry. Summation of the observations on the world geometry surfaces can provide a single value of the *information value* of that surface, and visual examination can present the *observation density* revealing the highest value location on that surface to place artifacts for interaction or desired observation. Conversely, this information can be used to identify good locations to hide artifacts in the virtual world. While this is a context-sensitive approach, it can be very useful to guide level design, art detail focus, and understand behavioral influence [3].

Data collection and analysis are a key part of modern game design and development. It can greatly assist in understanding how embedded virtual world information is used, misused, or simply ignored. It can guide attention resources in level design, information tagging, and many aspects of game creation toward issues of importance – improving the overall efficiency of the game creation process.

Key Principle: A wealth of additional information can be learned and used to improve game play and interactivity from logged data, and interactive visualization tools can significantly assist in this knowledge discovery.

3 Case Studies

The process of adding and deriving information to support game AI starts with model creation, it is integrated in level design, used in engine accessible services, dynamically expanded in-game, and augmented through additional discovery and analysis. The two projects in this section met project goals and were evaluated in their own unique ways; however, both depended on elements of information tagging and knowledge creation from the techniques presented here to be successful.

3.1 The Transfer Learning Project

The first year effort of the Defense Advanced Research Projects Agency (DARPA) Transfer Learning Program led by the ISLE Team (`www.isle.org`) used the University of Texas at Arlington's Urban Combat Testbed (a Quake3 total conversion mod) [21] to explore the mechanisms of learning in a source experience and applying that learned knowledge to a target experience where the difference between the source and target examined a particular type of learning (e.g., memorization, extrapolating, restructuring, abstracting, generalizing, and so forth). Sub-teams from Stanford University (ICARUS architecture agents), the University of Michigan (SOAR architecture agents), and the University of Texas at Austin (reinforcement learning agents) leveraged SSPS, DIAS, and an early version of PlayerViz to successfully show artificial transfer learning that in most cases exceeded that of humans under similar conditions [2]. This rare occurrence of using multiple academic, high-quality AI agent architectures in the same testbed was possible because of the information embedded and conveyed in the environmental objects.

Specifically, a navmesh similar to that shown in Fig. 4b was used for all agent navigation. World objects such as doors and crates were all annotated by Research Cyc constants and used to perform object and spatial reasoning using information provided in-game by SSPS and DIAS. The focus was primarily on solving spatial puzzles; so there was a heavy reliance on the information provided for the AI to solve problems by learning in one environment and then applying that solution to a similar problem, which sometimes came down to simple symbolic substitution (e.g., climb over create instead of a bush). However, none of this could have been accomplished without embedding information and providing dynamic updates.

3.2 DASSIEs

The DARPA-supported Dynamic Adaptable Super-Scalable Intelligent Entities (DASSIEs) Project addresses the problem of building an intuitive user interface for interactive character design. DASSIEs makes use of each of the elements described

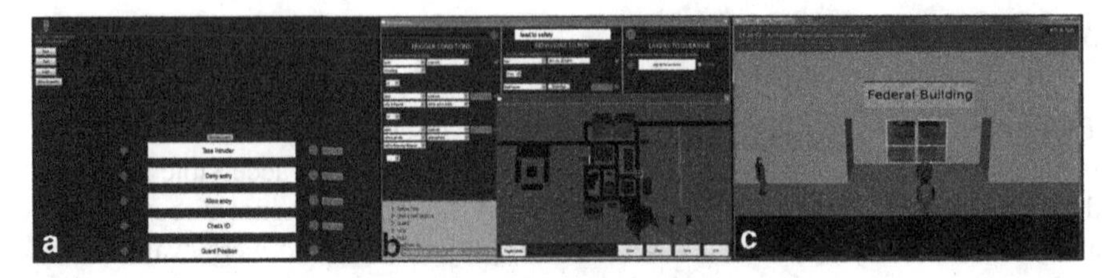

Fig. 7 The DASSIEs Project BehaviorShop v1.0 overall behavior layer builder, which is a graphical subsumption/behavior-based control specification tool, is shown in panel (**a**). The specific behavior layer builder including a graphical, spatial selector for the target environment is shown in panel (**b**). The BehaviorShop specified character is executed by the BEHAVEngine in our FI3RST (FIrst and 3rd-person Realtime Simulation Testbed) environment is shown in panel (**c**)

in Sect. 2 to support the BehaviorShop interface and the BEHAVEngine AI engine in multiple target game/simulation environments. Figure 7 shows the DASSIEs components in action.

BEHAVEngine is an AI engine for implementing AI characters using hierarchical behavior-based subsumption controllers [13]. Information from the simulation environment is received by a perceptual model, which augments the raw world state with additional information. This includes querying affordances to determine available actions. Part of this process can make queries to the character's working memory through the memory model. The memory model itself provides a store for character state information as well as a facility for making queries from relational ontologies (such as ConceptNet and ResearchCyc). Once the world state has been processed into *percepts*, this information flows into the subsumption controller. The subsumption controller makes decisions about what behaviors to run, and individual behaviors process the current set of percepts to generate appropriate actions. Behaviors may also query the perceptual and memory models, gaining access to working memory (probabilistic) affordance information, influence points, ontologies, and the SSPS/DIAS services. Finally, the behaviors send action requests to the action model, which can build commands for the simulation environment to execute actions, use affordances, and move through the world. The action model can also query the SSPS service to find paths through the environment.

BehaviorShop is a GUI for creating characters for BEHAVEngine [11]. It provides support for creating behavior for individuals and teams of characters in a modular fashion. Information about character behaviors is loaded from metadata provided by BEHAVEngine, while information about the simulation environment is received from the game/simulation environment from embedded information and information processed by the CGUL tools as described in Sect. 2. SSPS queries allow BehaviorShop to display maps of the environment, which allow users to more easily create spatial behaviors for characters. Object information can be queried from the game environment object database and includes information about affordances. Ontology information can also be used to specify the qualities of objects for use in the behaviors rather than binding specific object types or instances in the character definition.

DASSIEs highly leverages the work described in this chapter to make character behavior specification easy enough for the novice user to specify, while powerful enough to act intelligently in many roles in the game/simulation environment [5]. One of the key aspects of DASSIEs is that information flows from the world up. Having an information complete world with a built-in navigation mesh and structures for dynamic information storage and adjustment allows for BEHAVEngine to mostly just provide an architectural framework for behavior organization and firing. BEHAVEngine also provides a set of perception sensors, which convey much of the embedded information, and actuators, which provide a set of actions in which the characters can interact in the world. Many of the symbolic annotations from the world percepts are left ungrounded until specified for behavior inclusion in BehaviorShop. The base set of actions in BEHAVEngine is dictated by the affordance actions, but is often just scripted from a library of existing low-level actions and then grounded to the affordance language. BehaviorShop provides the character programming interface in a subsumption architecture, behavior-based control paradigm [1,17], which starts off as an empty shell and is then populated with action and perception choices passed from BEHAVEngine and the simulation environment. Most of the behavior specification is done through building piecewise English sentences or graphical selection/drawing (e.g., patrol routes). In this manner, the rich world information comes through the AI engine to the programming/configuration environment. Meaning is given to the symbolic information in BehaviorShop and executed in the environment using BEHAVEngine.

The key strength of DASSIEs and our approach to embedding information is that the subject matter expert (not a computer scientist) creating the models, environments, and context can effectively specify in a simple manner how they intend for those elements to be used in interaction or at a minimum how to be perceived. When propagated through a tool like BehaviorShop, they can then also more easily specify how the AI should behave in these environments. Thus, not only can we make smarter AI by providing more information, but we can also enlist the help of the non-programmer content creator to make better interactive AI.

4 Summary

Better AI in games and simulation comes at the expense of having more information in which to perceive, reason, and ultimately act upon. The current trend toward procedural and user-generated content as well as libraries of pre-made 3D assets unfortunately provides the means to continue to make information-poor virtual worlds. These environments are often difficult for AI to reason in because of the sparseness of information and lack of a designer/programmer to inject information. This chapter discusses the specific issues with the current trends in virtual world creation and provides some suggestive guidance on how to make and use information-rich worlds by embedding information into world objects and environments – building worlds with regard to the AI and physics engines as well as leveraging organizational tools to provide intuitive paradigms for reasoning about space. We forward the notion of having information complete objects and virtual worlds that makes them more amenable to AI. Techniques for handling the dynamic nature of these objects and environments are presented through probabilistic affordances and influence points to provide a complete notion of virtual world specification and use with regard to the AI, making light of the importance of embedded information and providing methods to handle embedded dynamic information. The importance of logging and analysis in understanding and improving virtual world construction and embedded information to move toward efficient and information rich environments is discussed, and we end with two case examples that leverage information-rich environments to achieve their advanced AI goals.

Acknowledgements Thanks are due to the following people, beyond those who are coauthors, who helped inspire, guide, and contribute to this body of work: Dongkyu Choi, Nick Gorski, Larry Holder, Nikhil Ketkar, Bharat Kondeti, Tolga Konik, John Laird, Pat Langley, Yaxin Liu, Cynthia Matuszek, Maheswar Nallacharu, Billy Nolen, Michael Ross, and Ashish Singh. This material is based on research sponsored by the US DARPA. The views and conclusions contained herein are those of the authors and should not be interpreted as necessarily representing the official policies or endorsements, either expressed or implied, of DARPA or the US Government.

References

1. Brooks, R.A.: A Robust Layered Control System For A Mobile Robot. In: IEEE Journal Of Robotics And Automation, vol. RA-2, pp. 14–23 (1986)
2. Cook, D.J., Holder, L.B., Youngblood, G.M.: Analysis of Human Transfer Learning Using a Real-Time Game Testbed. IEEE Transactions on Knowledge and Data Engineering (2007)
3. Dixit, P., Youngblood, G.M.: Optimal Information Placement in 3D Interactive Environments. In: Sandbox Symposium (2007)
4. Evans, R.: Implementation of Personality Traits in The Sims 3. Game Developer's Conference AI Summit (2009)
5. George Alexander, G. Michael Youngblood, Frederick W. P. Heckel, D. Hunter Hale, Nikhil S. Ketkar: Rapid Development of Intelligent Agents in First/third-person Training Simulations via Behavior-based Control. In: Proceedings of the 19th Behavior Representation in Modeling and Simulation Conference (BRIMS) (2010)
6. Gibson, J.J.: Perceiving, Acting, and Knowing, chap. The Theory of Affordances. Lawrence Erlbaum Associates (1977)
7. Hale, D., Youngblood, G., Ketkar, N.: Using Intelligent Agents to Build Navigation Meshes. In: Proceedings of the 23rd Florida Artificial Intelligence Research Society Conference (FLAIRS) (2010)
8. Hale, D.H., Youngblood, G.M.: Dynamic Updating of Navigation Meshes in Response to Changes in a Game World. In: Proceedings of the 22nd Florida Artificial Intelligence Research Society Conference (FLAIRS) (2009)
9. Hale, D.H., Youngblood, G.M.: Full 3D Spatial Decomposition for the Generation of Navigation Meshes. In: Proceedings of the Fifth Artificial Intelligence for Interactive Digital Entertainment Conference (AIIDE) (2009)
10. Hale, D.H., Youngblood, G.M., Dixit, P.N.: Automatically-generated Convex Region Decomposition for Real-time Spatial Agent Navigation in Virtual Worlds. In: Proceedings of the Fourth Artificial Intelligence for Interactive Digital Entertainment Conference (AIIDE) (2008)

11. Heckel, F.W.P., Youngblood, G.M., Hale, D.H.: BehaviorShop: An Intuitive Interface for Interactive Character Design. In: Proceedings of the Fifth Artificial Intelligence for Interactive Digital Entertainment Conference (AIIDE) (2009)
12. Heckel, F.W.P., Youngblood, G.M., Hale, D.H.: Influence Points for Tactical Information in Navigation Meshes. In: Proceedings of the 4th International Conference on Foundations of Digital Games (FDG), pp. 79–85. ACM (2009)
13. Heckel, F.W.P., Youngblood, G.M., Hale, D.H.: Making User-Defined Interactive Game Characters BEHAVE. In: Proceedings of the 22nd Florida Artificial Intelligence Research Society Conference (FLAIRS) (2009)
14. Heinze, C., Goss, S., Josefsson, T., Bennett, K., Waugh, S., Lloyd, I., Murray, G., Oldfield, J.: Interchanging Agents and Humans in Military Simulation. AI Magazine **23**(2), 37–47 (2002)
15. Hertel, S., Mehlhorn, K.: Fast Triangulation of the Plane with Respect to Simple Polygons. In: International Conference on Foundations of Computation Theory (1983)
16. Laird, J., van Lent, M.: Human-level ai's killer application: Interactive computer games. AI Magazine (22:15–25) (2001)
17. Mataric', M.J.: MIT Encyclopedia of Cognitive Science, chap. Behavior-Based Robotics, pp. 74–77. MIT Press (1999)
18. Priyesh N. Dixit, G. Michael Youngblood: Understanding Playtest Data through Visual Data Mining in Interactive 3D Environments. In: In the Proceedings of the 12th International Conference on Computer Games: AI, Animation, Mobile, Interactive Multimedia & Serious Games (CGAMES) (2008)
19. Tozour, P.: AI Game Programming Wisdom 2, chap. 2.1 Search Space Representations, pp. 85–102. Charles River Media (2004)
20. Youngblood, G.M., Dixit, P.: AI Game Programming Gems 7. Understanding Intelligence in Games Using Player Traces and Player Graphs. Charles River Media (2008)
21. Youngblood, G.M., Nolen, B., Ross, M., Holder, L.: Building Test Beds for AI with the Q3 Mod Base. In: Proceedings of the Second Artificial Intelligence for Interactive Digital Entertainment Conference (AIIDE) (2006)

Empowering Designers with Libraries of Self-Validated Query-Enabled Behaviour Trees

Gonzalo Flórez-Puga, David Llansó, Marco Antonio Gómez-Martín, Pedro P. Gómez-Martín, Belén Díaz-Agudo, and Pedro Antonio González-Calero

Abstract Building the behaviour for non-player characters (NPC) in a game is a collaborative effort between AI designers and programmers. Programmers provide to the designers with the building blocks for specifying behaviours in the game, and designers use some combination of state machines, scripting and visual languages to build complex behaviours by composing the basic pieces the programmers provide. *Behaviour trees* (BTs) are the technology of choice for AI programmers to build NPC behaviour. Although BTs can be naturally built using visual languages that require no programming, in general, they are considered too complex for being built by designers without a programming background. In this chapter, we propose a number of techniques for facilitating the collaborative work of behaviour design through BTs. We provide tools for creating and managing a library of reusable fragments of BTs, intended for both programmers and designers. Such library is accessed through retrieval mechanisms that also support the definition of query nodes in BTs that can be expanded at run-time. In order to harness such an expressive power in behaviour design, we also propose an extension to the component-based architecture that supports a number of sanity checks to validate BTs, both at design and run-time.

1 Introduction

Building the behaviour for non-player characters (NPC) in a game is a collaborative effort between AI designers and programmers. Programmers provide to the designers with the *building blocks* for specifying behaviour in the game, as a collection of parametrized systems, entity types and actions those entities may execute. Designers use some combination of state machines, scripting, visual languages and map editors to build complex behaviours by composing the basic pieces the programmers

---

G. Flórez-Puga (✉)
Complutense University of Madrid, Madrid, Spain
e-mail: gflorez@fdi.ucm.es

P.A. González-Calero and M.A. Gómez-Martín (eds.), *Artificial Intelligence for Computer Games*, DOI 10.1007/978-1-4419-8188-2_3,

provide. Just to give a hint about the magnitude of the task, developing a game such as Far Cry 2[1], according to [8], required an average number of 150 people (including testers) during 43 months, which results, making a conservative assumption of a 20% of designers, in 30 designers working for 3 years in creating game play content for a shooter.

Ideally, a detailed design document should serve as the specification contract between designers and programmers: before entering into production stage, it should be perfectly clear which building blocks the programmers should build and what building blocks the designers would count on for designing the game levels. However, in actual development, the design of the game usually becomes a moving target, with designers coming up with new requirements for programmers as new mechanics are explored. Furthermore, programmers overwhelmed by their current tasks can feel tempted to let designers use their dubious scripting skills to implement such additions, which, later on, will probably result in the programmer debugging a designer's script during crunch time.

A key problem in this process is that a good game designer may not have programming skills, but nevertheless what a designer is actually doing most of the time is building portions of a software system. A possible solution for this problem is to hire designers who know how to program, which actually some companies do (Double Fine fired the whole level design department in the mid of the development of Pshyconauts, and hired fresh college graduates from Computer Science departments to script the levels [3]). Another approach, also used in industry, is to let designers use visual languages that are supposed to facilitate the process, by hiding the formal syntax of the programming language, and controlling through a GUI the sentences that can be built with the visual language. UNREALKISMET, integrated in the Unreal Development Kit game editor [6], and FLOW-GRAPH EDITOR, integrated in the Sandbox Editor of CryENGINE 3 SDK [2], are two of such visual scripting tools that let designers model the gameplay of a level without touching a single line of code through some variation of data flow diagrams.

For AI programmers, according to the number of papers dedicated to the subject in the editions 3 and 4 of the AI Game Programming Wisdom book series [14, 15], behaviour trees (BTs) are the technology of choice for programming the AI of NPCs in different game genres. BTs have been proposed as an evolution for hierarchical finite state machines (HFSMs) intended to solve FSM scalability problems by emphasizing behaviour reuse [10]. In BTs instead of explicit transitions from one state to another, each node defines procedurally how to traverse its children. BTs are goal structures that represent how a high-level goal can be decomposed into lower level ones until reaching the leaves of the tree, which contain primitive goals that can be achieved by available actions. In this chapter, we propose a number of techniques for facilitating the collaboration between AI programmers and designers through the collaborative construction of BTs.

Although BTs can be naturally built using visual languages that require no programming, in general, they are considered too complex for being built by designers

[1] Far Cry 2 is a first person shooter developed by Ubisoft Montreal released on 2008.

without programming skills. The use of different levels of abstraction, implicit transitions and arbitrarily complex control structures for composite nodes make BTs as expressive as general purpose programming languages, and therefore not convenient for designers to use. Nevertheless, BTs have been successfully used by professional game designers in released commercial games, by focusing designers on building BTs for high-level strategic behaviour that relies on lower level reactive behaviour that programmers provide, typically also as BTs [22, 23]. Building upon this idea of BT fragments at different levels of abstraction, we provide tools for creating and managing a library of reusable fragments of BTs, intended for both programmers and designers. The library is equipped with an authoring tool that promotes to build new BTs by composing other BTs already in the library. Note that such a library supports collaborations between different roles: between programmers, who have an easy access to low-level BTs designed by other programmers, between designers accessing high-level BTs designed by other designers, and for designers to build high-level BTs reusing those that programmers designed. Considering the number of people involved and the duration of the process, as hinted above, having a principled way of accessing somebody else's BTs can become crucial to avoid a situation where BTs become a new form of spaghetti code that only its author, if anybody, dares to modify.

A library of reusable fragments of BTs requires a query language and a retrieval mechanism that returns BT fragments relevant for a given need. The query language that we propose is based on a declarative representation of the game world, a *domain model* that names and classifies the types of entities available in the game, along with their properties, available actions and goals. The same language will be used to annotate BT fragments with the intended goal, as well as the restrictions on the type of entities that can execute the BT or the parameter values it can receive.

The possibility of retrieving BT fragments from a library naturally leads to a second contribution of the work presented here. BTs can be extended to include *query nodes* that specify queries that will be executed at run time, resulting in the substitution of the query node with the retrieved BT fragment. This mechanism provides a controlled form of emergent behaviour, as well as an easy way to introduce variability in the responses of an NPC, and will also allow for high-level BTs to automatically incorporate new BT fragments as they are incorporated into the library.

Having designers build BT fragments with parameters and query nodes may easily result in unusable BTs. This may also be the case for BTs with query nodes even when designed by programmers, since BTs generated on the fly through this mechanism could be impossible to execute. Thus, to harness such an expressive power in behaviour design, we also propose an extension to the component-based architecture that supports a number of sanity checks to validate BTs, both at design and at run-time, through *reflective components* that are able to validate a given BT.

The rest of the chapter runs as follows. Section 2 presents the *BT* model that will be extended in later sections. Section 3 presents the mechanisms of a library of reusable BT fragments and shows how this naturally leads to extend BTs with query nodes. Section 4 presents the main ideas of a component-based architecture and how this can be extended to validate BTs. The chapter ends with a clarifying example and some conclusions.

2 Behaviour Trees

Finite state machines (FSMs) are the most used technology for AI on games, easy to understand, deterministic and fast. Designers are also used to them, and they can be defined using simple (even graphical) tools. Unfortunately when they are used to define complex behaviours, FSMs require more and more states that can become the FSM hard to control.

A way to scale up FSMs is to consider that a state can hide another FSM to decide its actions. Instead of having a flat set of states, they are arranged in different levels, creating an HFSM. Apart from adding more structure to the states, they ease the reuse of low-level FSM and provide different views of the HFSM depending on the detail they are observed, which facilitates their comprehension.

HFSMs expand the complexity of the AI of the NPCs that can be implemented with this technology, but, obviously, they also suffer of their own threshold that makes them too complex. Curiously, the bottleneck in the FSMs and HFSM scalability are not the states, but transitions. Transitions grow much faster than states, and they become uncontrollable sooner.

A way to overcome this problem is to completely remove transitions. The resulting structure is not a (H)FSM anymore but it is useful anyway. Without transitions, an AI of an NPC is defined using a "cloud of states", and a *procedural* way to choose which one is the active one. An AI of an NPC is not in a state anymore, but *executing a behaviour*. The selection mechanism that picks up the current behaviour hides the old nasty transitions and plays the role of a referee. It can use any information about the virtual environment to arbitrate between them.

This new scheme is enriched with a new ingredient: behaviours (the old states) *can end*. Although a behaviour could last many game cycles, eventually it could decide that it has finished its labour and a new behaviour selection should be triggered. Even better, behaviours can inform about the success or failure of their execution, information that enriches the decision-making process done by the selection mechanism choosing the new behaviour.

We can go even further considering the selection mechanism *as a behaviour* with sub-behaviours as *children*. With this fresh perspective, hierarchy comes to the surface: a behaviour could easily be implemented as a new low-level selection mechanism with its own sub-behaviours. This new decision structure is called *BT*. This step is similar to that taken when moving from FSMs to HFSMs. Now we have compound behaviours that are decomposed on sub-behaviours.

BTs can be drawn using a tree representation that could be confused with the FSM classical representation. Keep in mind that each edge in FSMs represents a *transition*, but BTs' edges represent parent–child relationships; an internal "decision node" chooses among all its children which one should be executed next; all "transitions" between behaviours are decided by those selector nodes, not by behaviours themselves as was done by states in a FSM.

Note that, depending on the context, nodes in a BT can be seen as states, behaviours or actions. In this context, "behaviour" is a synonym of (transitionless) "state", while "action" corresponds to a primitive behaviour that can only appear as a leaf in a BT.

The literature is full of proposals for different decision nodes; for the goals of this chapter, we only require three of them: sequences, static priority list and dynamic priority list. For all of them, the child order is important: children nodes are not a *set* of behaviours, but a *list*.

Sequences are simple composite behaviours that execute their children in the order they are defined. Keep in mind that behaviours *end*, so sequence nodes wait until the current active child ends *with success* to launch the next one. If any child *fails*, the sequence also immediately fails, throwing the problem up in the hierarchy. Sequences end with success when their last child does.

To introduce static and dynamic priority list, a new concept must be first presented. Children behaviours can be guarded by *conditions*, indicating when that child can be chosen. Keep in mind that these conditions are *not* preconditions, because a valid candidate child (which condition is true) could, after all, fail: true conditions do not guarantee the complete correct execution of the guarded behaviour.

With conditions in mind, a *static priority list* node evaluates its children conditions in order and activates the first one whose condition is true. The child order represents a *behaviour priority*, with the first child having a higher priority than the next ones. A *dynamic priority list* is similar, but it continuously *reevaluates conditions* of prior nodes to the active one and switches to a higher priority node whenever possible, as soon as its condition becomes true. In contrast to sequences, priority lists *fail* if all of their children fail. If any child ends successfully, the priority list also ends with *success*.

Although they are not important for this chapter, BTs usually provide with a second family of internal nodes known as *decorators*. Decorators have only one child, and they add or modify the original child behaviour. Examples of decorators are *control modifiers* (negating the child result, or forcing a concrete one) or filters (repeating the child behaviour while it succeeds, avoiding it to be fired too often using a timer, etc.). Decorators bring into BTs the expressive power of a general purpose programming language [13].

Apart from the lack of transitions, other crucial aspect to overcome the scalability problem in FSMs is considering nodes as *behaviours* instead of *states*. This new point of view introduces the idea of a *goal* for every behaviour and, with this vision, design is simpler because hierarchy let designers think in terms of goals and subgoals instead of states and substates. Most actions have a primary goal along with a number of additional goals that depend on the action context [24]. For example, the primary goal of the action "move-to" is to change location from x to y, but in an urban fight scenario we can be moving to get under cover from enemy fire or to assist a fallen comrade. Having actions focus only on their primary goal can sometimes lead to unintelligent behaviour. For example, if an agent is moving to a destination and is attacked, it will continue to move, even when it would be totally destroyed by doing so. Instead of adding conditional statements to every action that specify all the exceptions to normal behaviour, we can handle multiple goals and make them part of a hierarchy, which prioritizes goals higher up in the hierarchy, i.e. staying alive is more important than moving to point *y*; so if some condition higher up in the BT becomes activated for self protection, the whole branch being executed can be pruned.

Hierarchy also supports reusability, because a BT fragment can be seen as a black box that provides a specific behaviour that can be attached to more complex BTs as a child. Throughout the game production, more and more behaviours (general and enough reusable BTs) will be available for the designers' team, saving time from reinventing the wheel.

2.1 A Domain Model for Behaviour Trees

For reusability becoming true, reusable BTs should allow some kind of parametrization. For example, designers may build a BT for an enemy that attacks using an available weapon and picks an item up afterwards. Although the concrete weapon and item could be hard-coded in the BT, this spoils nearly all opportunity for reusing it; so an elaborate mechanism to specify parameters should be available. The system should be good enough to let parameters be bound both in design and runtime, depending on the circumstances.

Keep in mind that both FSMs and BTs are *static* structures used to model *NPC AI*. In runtime, the same FSM could be used for multiple NPC simultaneously, each of them storing the current state and other information needed to "run" the FSM. Something similar occurs for BTs, where each NPC should keep track which *behaviours* are activated, which ones have failed and so on. For parameter passing between nodes, the NPC runtime structure is enriched with an "execution context" (or *blackboard*) specific for each NPC, where behaviours read information (attribute–value pairs) to be used in the decision-making process using the guards (conditions). The set of attributes in the context is the portion of the game state that can be accessed by the NPC. Values will be specified by designers during development (e.g. to force an NPC to pick up a *concrete* weapon), or written by some actions (leaf behaviours) in runtime (e.g. the treasure found by a search behaviour).

In order to be able to reason with BTs independently of the underlying game engine, we need to model the context and parameter passing mechanisms. Furthermore, we need to specify the collection of goals and the restrictions on the type of entities that can execute the BT or the parameter values it can receive. We propose the use of ontologies to represent both the knowledge and the entities. Ontologies are a standard mechanism for knowledge representation, based on conceptual hierarchies, defined using the *is-a* relation where abstract concepts are located on the top of the taxonomy, while specific concepts are located in its leaves.

To model the knowledge on our domain we use a *behaviour ontology*, which provides different classes used to categorize the behaviours in terms of the *goals* they fulfil. In the ontology, we can find behaviour classes like *Attack Behaviours*, *Defend Behaviours* or *Resource Gathering Behaviours*. Each class can have several instances that represent the different behaviours for that goal. For instance,

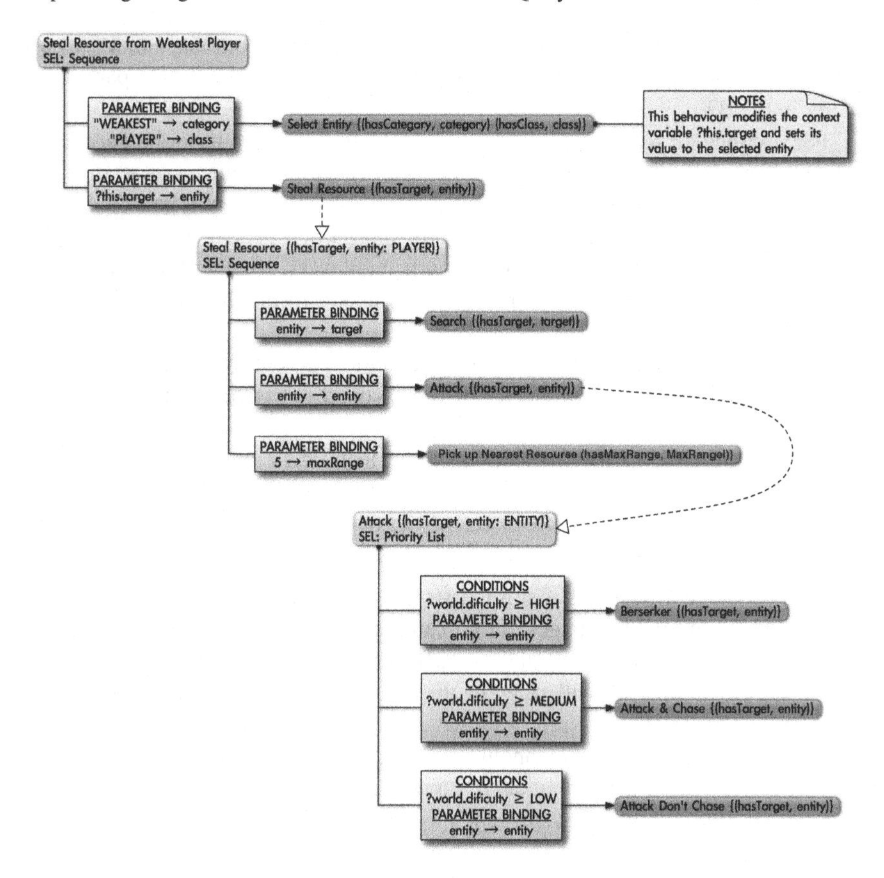

Fig. 1 Behaviour tree (BT) for *steal resource from weakest player*

in the class *Resource Gathering Behaviours*, we can find `Steal Resource From Weakest Player` (Fig. 1) or in the class *Attack Behaviours* we can find behaviours like `Long Range Stealth Attack` or `Hand To Hand Stealth Attack`.

To classify the entities that form the context in which behaviours are executed, we use an *entity ontology*. In the top of the entity ontology, we can find, for instance, classes like *Alive* that represents the alive creatures. Going down through the ontology, we will have subclasses like *Monster* and *Player* that are alive creatures. *Player*, in turn, subsumes the *Human* and *Computer* categories that respectively represent the player's avatar and an AI-controlled avatar.

Additionally, a set of relations exists between behaviours and entities. These relations are used to express the restrictions on the parameters of the behaviours.

Parameters are referenced in two places in the BT:

- The set of parameters that will be used in the BT is declared in the root of the tree. Each declaration consists in three elements: the relation between the behaviour and the entity in the parameter, the class from the entity ontology that will be the parent of the entity and the name that will be used to reference the parameter later in the BT.

For instance, in the `Steal Resource` BT in Fig. 1, we have the parameter declaration `(hasTarget, entity: PLAYER)`. This means that an input parameter is declared for the relation `hasTarget`. The entity type of the input parameter is `PLAYER`, which means that the *target* of this behaviour can only be a PLAYER (resources can only be stolen from players, whether they are human or AI controlled). To reference this input parameter inside the BT, the identifier `entity` should be used.

- In the invocation of other BTs or leaf behaviours, parts of the execution context are bound to the input parameters of the invoked behaviour in the parameter passing mechanism.

  This is the case of the invocation of the leaf `Search` by the BT `Steal Resource From Weakest Player`. In this case, the value of the parameter `entity` from `Steal Resource From Weakest Player` is bound to the input parameter `target` of `Search`.

The NPC context provides a storage structure similar to that found in object-oriented programming languages. For example, `?this` will refer to the NPC executing the BT (with information such as `?this.health` or `?this.aggressive`), `?world` will refer to the virtual environment state (`?world.time`) and `?target` will refer to the game entity target for the behaviour (`?target.distance`). As a conclusion, NPC context provides a way to consult the *game state*, both of the virtual environment and the NPC state itself.

Using this notation, we can represent a tree such as the one shown in Fig. 1.

3 A Library of Reusable Behaviour Trees

One of the main advantages of using BTs for the AI design is the reusability they provide. The main reusability components are basic actions provided by programmers, but BTs combining several nodes could also become reusable behaviours to be chained into more complex BTs. For example, a programmer could create a `StealthWalking` BT using simple actions that look for dark zones and walk through them. Once it is available, other designers could use it to create behaviours such as `SurpriseAttack` or `Spy`.

BT reusability is possible because of two features that are common in most of everyday videogames. First of all, modularity in behaviours: complex behaviours can be decomposed into simpler behaviours that are somehow combined. Second, simpler behaviours tend to recur within complex behaviours of the same game, or even in different games of the same genre. For instance, in an action game, a `Hand to hand attack` could be a complex behaviour that is composed of two simpler behaviours like `Go to (enemy)` and `Attack with knife`; on the other hand, `Long range attack` could be composed of `Go to (cover)` and `Shoot ray gun`. Both features are useful to build new complex behaviours based on simple behaviours as the reusable building blocks.

Programmers and designers should keep an eye on BT reusability in two aspects. They should create BTs trying to make them general enough to be later reused. And, at the same time, they should try to reuse previously made BTs, instead of reinventing the wheel creating the same basic behaviours again and again. This is quite important because although BTs make easier the creation of behaviours for NPCs, it still take a lot of time to wire them up because of the large number of behaviours that can be involved in the process (Halo 2 had an average of 60 different behaviours arranged in 4 layers [10]).

To assist game designers in the creation and edition of BTs, we have developed the *eCo Behaviour Editor*. The *eCo* editor is an authoring tool that provides the users with a graphical interface which allows them to manually create or modify behaviours just by "drawing" them. It includes tools for loading, saving and importing the behaviours from disk, drawing and erasing nodes and edges from the trees, and specifying their content. Once the behaviour is complete, it is possible to use the included code generation tool to generate the source code corresponding to the behaviour.

Nevertheless, the more outstanding feature of the *eCo* editor is BT reusability. Every manually designed behaviour is *stored and indexed* in a *database* that allows easy *BT retrieval* of previously stored behaviours. We use techniques imported from the case base reasoning (CBR) area, where data (cases) are stored in such a way that search becomes more than only matching.

CBR is based on the intuition that new problems are often similar to previously encountered problems, and therefore that past solutions may be reused, directly or through adaptation, in other situations. CBR systems typically apply retrieval and matching algorithms to a case base of past problem–solution pairs. Another very important feature of CBR is its coupling to learning. A strong effort has been done in the CBR community to solve the problems of similarity and adaptation in different contexts, with different approaches to case representation, organization and storage, and amount of knowledge, from knowledge intensive to data intensive approaches.

CBR is specially well suited to deal with the modularity and reuse properties of the behaviours; it assists the user in the reuse of behaviours by allowing her to query a case base. Each case of the case base represents a behaviour. By means of these queries, the user can make an approximate retrieval of behaviours previously created, which will have similar characteristics and satisfy some conditions. The retrieved behaviours can be reused, modified and combined to get the required behaviours.

Although the more important component of each case is the BT itself (the behaviour that want to be retrieved), they also store *metainformation* that is used in the search process. We use XML files to store all this information, which is defined by the following attributes:

1. *Header*: Includes the *case number*, used to identify the case in the case base, and a *textual description* that describes in natural language the behaviour represented by the case.

2. *Goals*: This attribute enumerates the list of goals from the *behaviour ontology* satisfied by this behaviour.
3. *Parameters*: This attribute is the set of parameters received by the behaviour (e.g. the enemy to attack or the weapon to use), along with the restrictions of type of each one of them. The type is built from the classes in the ontology which an individual belongs to.
4. *Descriptors*: This attribute is a set of restrictions declared over the game state (context variables such as `?this`, `?target` or `?world` mentioned previously). The values of the descriptors can be either symbolic or numeric. The descriptors specify under which circumstances of the game state is appropriate to run the behaviour.

As an example, Table 1 shows the set of behaviours that satisfy the goal *Attack*.

When a designer creates a new BT, she must enrich it with all this information. Although this could be seen as tedious and useless, they could be used it later while retrieving previously stored BTs to be mixed with new ones.

We distinguish between two types of queries: functionality-based queries and structure-based queries. In the former, the user provides a set of *descriptors* to specify the desired functionality of the searched behaviour. In the latter, a behaviour is retrieved whose composition of nodes and edges is similar to the one specified in the query.

3.1 Functionality-Based Retrieval

The most common usage of the CBR system in the editor is when the user wants to obtain a behaviour similar to a query in terms of its functionality. The functionality is expressed by means of a set of descriptors regarding the game state.

The *eCo* editor provides a query form, shown in Fig. 2, for the user to enter the parameters of the query. The attributes that form a query are:

1. *Goals*: Goals of the behaviour ontology that must fulfill the retrieved behaviour. The class of the goal can be selected in the tree on the left side of the query form, which shows the behaviours taxonomy. The query may only retrieve behaviours for the selected class or any of its subclasses.
2. *Parameters*: Restrictions on the type of the input parameters of the retrieved behaviours. For example, `weapon` should be a `firearm`.
3. *Descriptors*: A set of restrictions declared over the game state that describe the behaviour to be retrieved.
4. *Weights*: The weight of each descriptor in the final similarity calculation.
5. *Textual description*: A natural language description of the behaviour that will be compared with the description in the header of the cases. The textual description allows the user to fine-tune the search.
6. *Cases retrieved*: The maximum number of behaviours the user wants to be retrieved.

Table 1 Behaviours that satisfy the goal *Attack*

Case	Parameters	Goals	Descriptors
C_1	Hand-to-hand stealth attack (hasTarget, entity: ALIVE)	Attack	?target.distance ≤ MEDIUM ?this.personality = STEALTHY ?this.defensive ≥ MEDIUM ?this.health ≤ MEDIUM ?this.underAttack = LOW ?world.time = NIGHT
	Tries to approach an enemy without being noticed and attacks him using a close range, stealthy weapon. The entity executing it must remain undetected for the behaviour to be effective		
C_2	Long range stealth attack (hasTarget, entity: ALIVE)	Attack	?target.distance ≥ MEDIUM ?this.personality = STEALTHY ?this.defensive ≥ MEDIUM ?this.health ≤ MEDIUM ?this.underAttack = LOW ?world.time = NIGHT
	Looks for cover in the surroundings and attacks the enemy with a stealthy weapon. The entity executing it must remain undetected for the behaviour to be effective		
C_3	Berserker (hasTarget, entity: ALIVE)	Attack	?target.distance = MEDIUM ?this.personality = BRUTE ?this.aggressive = HIGH ?this.health = HIGH
	Attacks an entity with the most powerful weapon available and without caring about own safety. This behaviour is used for very aggressive entities. A defensive entity will not show this behaviour		
C_4	Grenade attack (hasTarget, entity: ENTITY)	Attack	?target.distance = HIGH ?this.personality = BRUTE ?this.aggressive ≥ MEDIUM
	Throws a grenade to an enemy and takes cover to avoid being affected by the explosion		
C_5	Elusive attack (hasTarget, entity: ALIVE)	Attack	?this.personality = TIMID ?this.aggressive ≤ MEDIUM ?this.defensive ≥ MEDIUM ?this.health ≤ MEDIUM ?this.underAttack ≥ MEDIUM
	Approaches the enemy and shoots him while trying to cover behind the objects in the game world and zigzags to avoid being hit. It is a defensive behaviour useful when the entity is being attacked or when the health is low		

The execution of the query goes as follows. First of all, the cases for the *Goal* specified in the query are retrieved. The similarity with the remaining cases is considered 0. If the user has specified any restrictions on the *Parameters*, they are checked. Any candidate who does not satisfy the *Parameters* restrictions is excluded from the candidate set (again, its similarity is 0).

Fig. 2 Retrieval interface

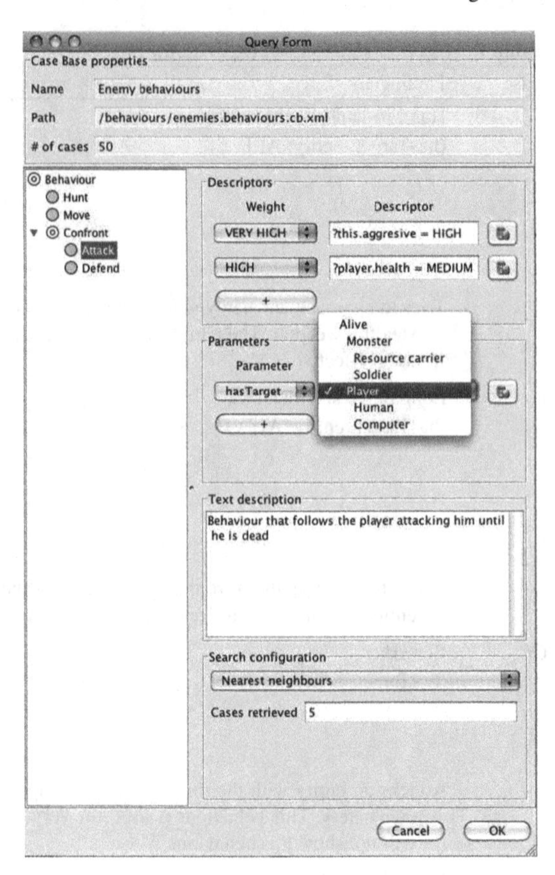

Then, the attributes of the query are compared to the attributes describing the BTs in the case base using a similarity function. Given a query, Q, and a case from the case base, C, the similarity value is obtained as follows:

$$\mathrm{sim}(Q,C) = \begin{cases} \bullet \text{ The class of C does not belong to the goals of Q} \Rightarrow 0 \\ \bullet \text{ The restrictions on parameters in Q do not hold in C} \Rightarrow 0 \\ \bullet \text{ otherwise} \Rightarrow \mathrm{sim}_{atr}(Q,C) \end{cases}$$

$$\mathrm{sim}_{atr}(Q,C) = \sum_{d \in D(Q,C)} w_d \cdot \mathrm{sim}_{loc}(Q_d, C_d)$$

$$D(Q,C) = Q.\mathrm{descriptors} \cap C.\mathrm{descriptors}$$

$$\mathrm{sim}_{loc}(Q_d, C_d) = 1 - \frac{|Q_d.\mathrm{value} - C_d.\mathrm{value}|}{\mathrm{size}_d}$$

$D(Q,C)$ is the intersection of the sets of descriptors of Q and C and size_d is the size of the interval of valid values for a descriptor d. Each w_d is the weight corresponding to the descriptor d, normalized so that the sum of all the w_d is 1.

To obtain the global similarity value between each of the cases and the query, the weighted similarity of the *Descriptors* is aggregated with the similarity due to the *Textual description* of each behaviour. Using a string similarity measure, the *Textual description* of the query is compared to the description in the *Header* of the case.

Finally, the candidates are sorted by their similarity value and the most similar ones to the query are retrieved.

3.2 Structure-Based Retrieval

In some circumstances, the behaviour designer knows the general structure of the BT (i.e. the distribution of the nodes and their generic functionality). In these situations, it would be easier and faster for the designer if he could "sketch" the tree and let the editor find a similar one in the case base.

The user can draw a tree with empty nodes (a tree pattern) and let the system find a similar one with all nodes defined. But, by entering these data alone, the retrieved BT would be similar to the query only in terms of its shape. The behaviour it implements could be any. Hence, we need to allow the behaviour designer to point out the desired functionality of the retrieved tree, and then compare the desired functionality with the functionality implemented in the nodes of the trees in the case base.

The functionality of the drawn nodes is expressed by linking each node to a *Functionality Query* that the user must build to express the desired behaviour that should be contained in the node. The linked functionality queries are compared to the descriptors in the nodes of the behaviours in the case base during the query process.

Keep in mind that a *functionality-based query* (previous section) could be specified using one of these new structure-based queries, drawing just a root node with no children. In that sense, we can see structure-based retrieval as an additional refinement search step, where the designer wants to impose some structural restrictions to the children nodes. Retrieval compares the sketched tree with those BTs in the case base, using any of the existing techniques in the literature for comparing ordered trees (like [18, 19, 21]).

Our approach to these structure-based queries is to use the drawing facilities of the editor to "draw" the BT pattern, and then assign functionality-based queries to the nodes, which will show the functionality of each node. Figure 3 shows the query editor for the structure-based queries. In the left pane the user can draw a behaviour pattern and in the right pane he or she can specify the desired functionality of the retrieved behaviour by entering a functionality query. Additionally, each node can be linked to another functionality query, as we have already mentioned, to tune up the search.

Further explanations regarding functionality- and structure-based retrieval can be found in [4].

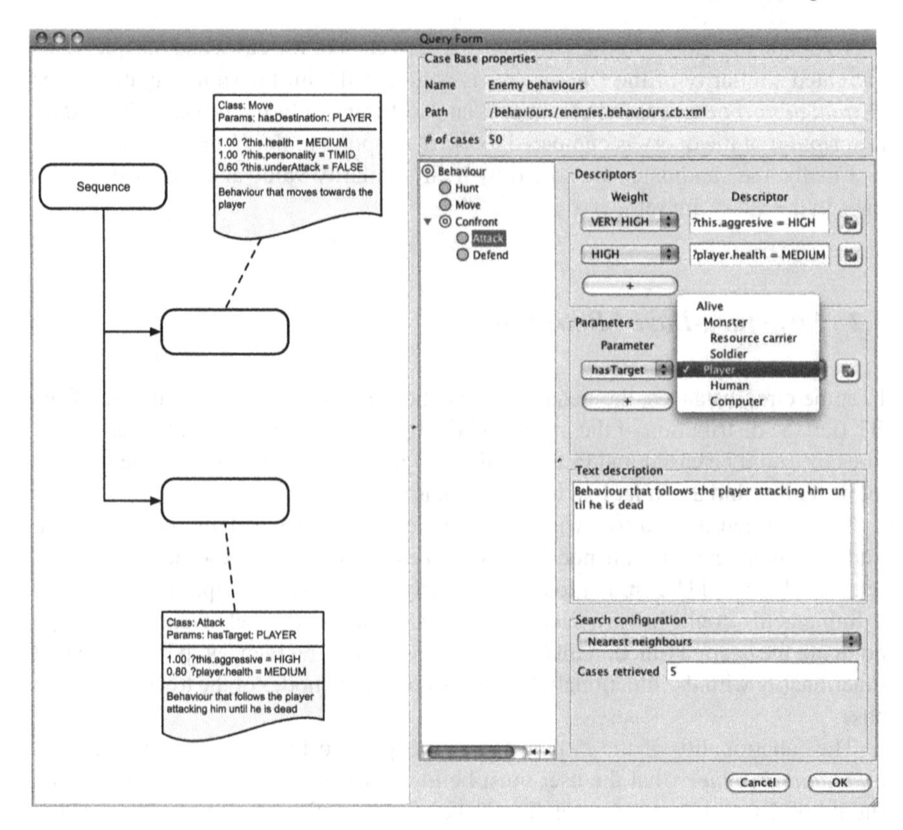

Fig. 3 Structural queries

3.3 Query Nodes at Run Time

Reusability and modularity are important advantages of using BTs. Each BT represents an abstraction that can be reused as a composing piece of other BTs. Different BTs are created independently during the game design phase and they can be assembled as pieces of other existing BTs. The collection of game BTs includes different ways of solving a certain goal, e.g. different ways of getting food or stealth walking.

The search facilities included in our *eCo* editor described in the previous section provides *static reuse*: once the behaviour designer has chosen a suitable BT provided by the query, it is tied to the new BT been created. However, throughout the game development, both programmers and designers add more and more reusable BTs that could have been also suitable (even better) for those searches done previously. Then, to make the process consistent and useful, it is important to review the pre-existing BTs that include a certain goal to check whether it is convenient to assemble newer BTs (representing new ways to solve certain goals). This consistency checking process generates an extra effort that is sometimes skipped. That means that the behaviours added in the late design phases are not taken into account by the behaviours that were included in the early design phases.

To address this problem we propose a dynamic approach where the CBR system is queried at run time to find the most appropriate behaviour from a case base of implemented behaviours using BTs. The CBR processes work always with an up-to-date behaviour case base that allows retrieving the most convenient behaviour according to a certain query using the whole collection of designed behaviours and avoiding the extra cost of pre-checking its adequacy with newer behaviours.

Keep in mind that the reusing possibilities described in the previous section were an extra functionality provided by the *eCo* editor, which do not require runtime infrastructure in the BT framework. However, runtime queries require a new BT node, called *query node*, that stores the query attributes specified in design time and makes the *BT retrieval* at runtime.

The attributes that describe these queries are the same ones used in the queries at design time (Sect. 3), adding a new *requery* field. Once the behaviour has been retrieved and is running, there may occur changes in the game state that would make another behaviour more suitable for the current situation. Using the *Requery* parameters, we can specify the conditions or changes in the game state that should make the system repeat the query. Note that, although the query is done again, the results can be the same. In that case, the behaviour being executed is not restarted.

Although we have defended the advantage of runtime queries because they use all the available BTs, they provide an even more important benefit: they can use the *current world state* to select the more suitable behaviour. Parameters can now refer to the complete game state (`?this`, `?world` and `?target`), not only to static restrictions on the input parameters.

The retrieval process is very similar to the one explained for the functionality based retrieval. The main difference is that the values of the descriptors are not specified in the query. In this case, the query specifies the relevant descriptors and the values are taken from the game state at runtime, at the instant of time that the query is run.

Figure 4 shows an example of a *query node* that retrieves an Attack behaviour.

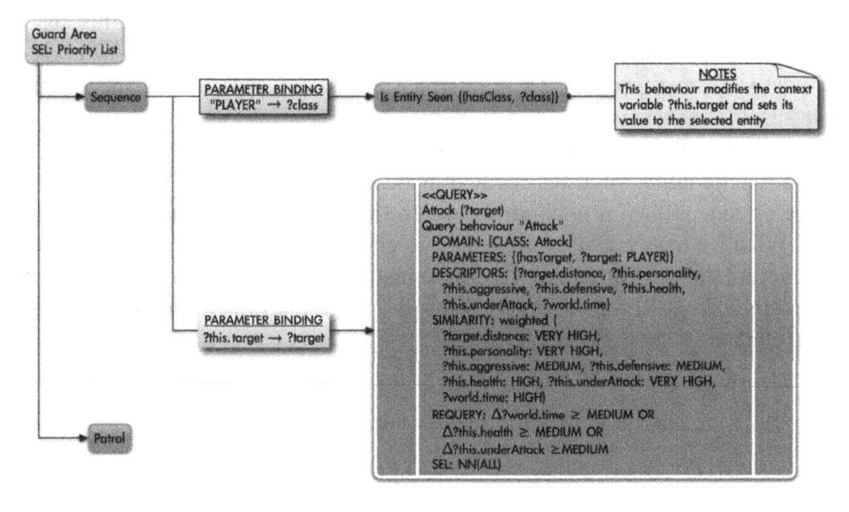

Fig. 4 Guard area behaviour

4 Reflective Components

Although runtime search of BTs provides a lot of advantages against the static search at design time, they can become quite dangerous because the retrieved BT could not fit the NPC features. For example, an NPC could query for a behaviour to run away from the player and receive a behaviour that uses a nearby car. At runtime, the system should check whether the retrieved BT is suitable for the target NPC answering questions such as whether the NPC can drive.

Before explaining our proposal to solve this problem, we need to introduce some implementation details about how game entities are usually coded. The runtime object-management system is in general an important part of a videogame, and creating this piece of code takes a great amount of time. To mention just two examples, a mature game such as Half-Life dated at 1999 has more than 65,000 non-empty no-comments lines of code on that module, while Far Cry at 2004 exceeds 95,000 lines of C/C++ code[2] even though the majority of the module was actually written in LUA [9].

This reveals that when creating this module we should try to design it to promote reusability in the sense that every single piece of code general enough to be used on a different title should be reused.

When we inspect how this module is usually coded, we find that it was traditionally based on an inheritance hierarchy, where all different kinds of entities derived from the same base class often called `CEntity`. Some of the consequences of this extensive use of class inheritance were an increase in the compilation time [11], a code base difficult to understand and big base classes. To mention just two examples, the base class of Half-Life 1 had 87 methods and 20 public attributes while Sims 1 ended up with more than 100 methods. The consequence is the well-known *fragile base class problem* [17].

Due to all these problems, today developers tend to use a different approach, the so-called component-based systems [1, 7, 16, 20]. Instead of having entities of a concrete class which define their exact behaviour, now each entity is just a *component container* where every functionality, skill or ability that the entity has is implemented by a component. From the developer point of view, every component inherits from a specific class or interface (called, for example, `IComponent`), while an entity becomes just a list of `IComponents`.

As the components are now generic objects with a common interface independent of their functionality, the usual method invocation is not enough. We cannot have a piece of code calling a method like `moveTo()`, because no such method even exists. What we have now is a *component* (a class called for example `MoveTo` that inherits from the previous `IComponent`) that is able to move the entity from one point to another; however, externally this is just an `IComponent` indistinguishable from the other.

[2] Lines of code (LOC) obtained using SLOCCount by David A. Wheeler.

The communication is therefore performed in a different way, using message passing. The `IComponent` is viewed as a *communication port* that is able to receive and process messages. A message is just a piece of data with an identification and some optional parameters (the implementation may vary from a plain `struct` with generic fields used in different ways depending on the type of message, to a base class such as `CMessage` and a hierarchy of messages). Components have a method like `handleMessage()` that is called externally to send the piece of information to it; depending on the concrete component, the message will be ignored or processed accordingly. In this scenario, entities play the role of the broadcaster of messages. Both, internal components and external modules may send messages to the entity that are automatically distributed among all of its components. The message types that a component can process usually correspond with basic entity-actions. Consequently, an entity is able to execute as many actions as the sum of messages its components can process. For instance, when the AI component (which provides the entity with the ability to *think*) wants to move the entity from one point to another, it *sends* a `MoveTo` message to *all* the components of its entity. The component that implements the ability of movement (`MoveTo` component in our previous example) intercepts the message, calculates the path to be followed and emits periodically `UpdatePosition` messages to notify other components (graphical and physical among others) the change of the position.

As entities are now just a list of components, the concrete components (or abilities) that constitute them may be specified in an external file (usually known as *blueprint*) that is processed in execution time. This approach eases the creation of new kind of entities, because it does not require any development task but just the selection of the different skills we want our new entity to have from a set of components.

The approach also fosters the reuse of the components in other projects. As the responsibility of every component is neatly defined and it is in charge of just a small set of tasks, most of them are general enough to be useful in other applications.

In order to allow fine-grained adjustment of the behaviour (or skills) of different entities, their definition may also set the values of different attributes that components use as parameters of their behaviours. For instance, the component that provides the entity with the ability of picking up objects may use an attribute that specify the strength of the entity.

Keep in mind that entities construction in runtime is now *generic* due to the *blueprint* file described previously. Therefore, the concrete parameter values (such as the strength of each NPC race) must be also provided as data instead of being hard coded in source. This information is also provided in an external file, known as *archetypes*, containing the default values for each parameter of each entity in the *blueprint*. Map files for game levels will have the opportunity to override the default archetype values for some concrete entities, providing, for example, more strength than the default one to a specific NPC.

As an example, Fig. 5 presents a `Patrol Soldier` entity built by components. This figure contains parts of the both mentioned entity descriptions files where the blueprints file reflects the abilities of the entity as a collection

```
<!– blueprints file –>
<blueprints>
   <entity type="Patrol Soldier">
      <component type="AnimatedGraphic"/>
      <component type="Physic"/>
      <component type="BTExecuter"/>
      <component type="MoveTo"/>
      <component type="ShootTo"/>
      <component type="HandToHandAttack"/>
      <component type="Skills"/>
   </entity>
   ...
</blueprints>
...
<!– archetypes file –>
<archetypes>
   <entity type="Patrol Soldier">
      <attrib name = "life" value = "100"/>
      <attrib name = "speed" value = "1.5"/>
      <attrib name = "arm" value = "gun"/>
      <attrib name = "arm" value = "rifle"/>
      <attrib name = "model" value = "patrol_soldier.n2"/>
      ...
   </entity>
   ...
</archetypes>
```

Fig. 5 `Patrol Soldier` entity built by components

of components and the archetypes file displays the attribute–value pairs that makes the fine-grained data-oriented entity description possible. The blueprints shows that a `Patrol Soldier` can be rendered and animated, can collide with other physic entities, execute BTs, walk from one place to another, etc., while the archetypes file sets the entity attributes to their default values.

Note that both entity description files just add information to the *entity ontology* described in Sect. 2.1, where the `Patrol Soldier` would be a specialization of the *Computer* category that represents an AI-controlled avatar. So entities in the blueprint and archetypes files must fit with entities described there.

It is important to stress that our entity ontology just simplify the entity distribution, in a high level, through *is-a* relations, but these relations *do not* involve that a child concept has all the abilities that its parent ontology concept has. Entity ontologies are excellent mechanisms to take high-level decisions, but their *is-a* relation is not a good idea to implement low-level details in big projects as it has been exposed in this section. That is the reason why the entity ontology is not translated in hierarchy classes when implementing our games in a programming language such as C++ but in entities built by components.

The reusability that components give us comes at a price, though. As the entity definition is made from text files, the consistence of the created entity class is not guaranteed. Prior the use of components, when new entity class was developed completely in a programming language such as C++, the compiler itself checked whether the new class was complete before allowing programmers to create an object of it. Therefore, an entity with the ability of, say, walk to a location was *always* able to set the `walk` animation; otherwise the `setAnimation()` method invocation would not have compile.

When the declaration of entities becomes the addition of a set of lines in a text file, developer may forget to provide the entity class with some ability that is needed by other components. Following with the previous example and noticing the `Patrol Soldier` entity blueprint (Fig. 5), if the entity has the ability to walk from one point to another (which implies the blueprint file states that the entity possesses a particular component such as the `MoveTo` component), it should also be able to change the animation presented on the screen (possessing another component such as the `AnimatedGraphic` component) because the `MoveTo` component just sends a `SetAnimation` message and update regularly the entity position sending `SetPosition` messages.

Our solution to this problem is what we call *reflective components* [12]. This technique consists in enhancing components with some methods that allow us to check, at design time and even at run time, whether an entity is able to perform an action (and therefore has that particular ability).

During runtime, components that are related to the behaviour of the entity (*AI components*) such as those that manage BTs (`BTExecuter` component in Fig. 5), send messages to the entity they belong to order which actions must be executed. This is due to AI components that do not have the ability of executing these primitive entity actions because they only perform the decision-making process. In that sense, an entity with just the `BTExecuter` component is not complete, because it is not able to actually execute the tasks that the AI selects.

Because now entities are specified in terms of their components, and that a component can be seen as an ability that an entity has, it makes sense therefore to try to identify the failures related to the inherent nature of the entities using such a description. The easy (and naive) approach is to make direct associations between basic actions (or messages that represent them) and components that are capable of executing these actions (process these messages).

Nevertheless, this approach would not be enough. Sometimes a component could not be able to carry out an action, although it has the ability to do it, either because it needs the collaboration of other components, which may not be in the entity, or because the component cannot correctly execute the action with the parameters associated with this action.

Let us imagine a situation where the `BTExecuter` sent a `MoveTo` message to makes the entity walks. The only existence of the `MoveTo` component would not assure the correct execution of the action since the `MoveTo` component would send `setAnimation` and `SetPosition` that other components should process. In the same way, the existence of a `AnimatedGraphic` component would not assure the correct execution of a `SetAnimation` message because the 3D model associated with the component could not have the given animation.

In order to manage both kinds of errors and with the purpose of giving a fine-grained approach, the methods we propose to enhance the component-based systems will accept messages that encapsulate actions to ask whether they are able to handle a concrete message according to their configuration. So, we shall query them using the same messages that BT actions (or other kind of *AI systems*) generate during the game execution. Then, if any component needed the collaboration of other components, it would only have to query the entity it belongs with the same message that it would generate during runtime and finally the component would return if the collaboration succeeded. Furthermore, as messages and components are parametrized, the new check methods can carry out fine-grained approach using them in the association process.

So, to check whether a full BT can be carried out by an entity, the `BTExecuter` component, for every action of its behaviour, just has to query to their entity components if some of them are able to execute it. Note that this is general enough. In games where a task may be performed using different methods, each entity capable of performing that task will be provided with the component that executes it using a particular method. As the behaviour component queries for the *ability* of executing the *task* instead of asking for the particular component that implements a method, the consistency check will work.

Again, we have a coarse-grain approach, though, not due to the reflective components but due to the BT action iteration. Just iterating over the list of actions of the BT is not precise enough since, in this way, the system would only validate or invalidate associations between BTs and entities; but if the system invalidated an association, it would not locate where and why this association was invalidated, so it cannot be fixed easily.

Therefore, a fine-grained approach should locate which branches of the BT were not able to be carried out by the entity and which node and the reason that made it crash. Bearing in mind that a BT may have different decision nodes (Sect. 2) and the chosen children to be evaluated depends on them, different kinds of nodes have to be evaluated by different methods.

As its name denotes, a sequence represents a chain of behaviours. Thus, to validate a sequence, all its children nodes must have been validated with the entity before. Therefore, if there was one node of the sequence that was not validated, the whole sequence would be invalidated knowing why and where the problem would be.

But both static and dynamic priority lists represent a behaviour that chooses between different ways of resolving a problem. So only one of the child nodes would be executed during the game rather than in previous example, in which all the children would be executed (sometimes more than one child is executed, but only if there are more children available and child guards changes during the selector execution). As a result of this, a fault detected in a child of the selector was less critical than faults detected in a child of sequences. This is because there were probably another choice (other child) selectable by the selector. Therefore, we could call these faults as warnings, instead of failures, if the child node that fails has at least one other brother node that has been validated.

So, although interested readers are referred to [12] for more details, to summarize, to validate a BT with an entity, the BTExecutor would try to validate the root node of the BT with the entity it belongs and this would be recursively spread to all the nodes of the tree. Finally, the leaves of the tree, which contains the final actions, would be checked with the entity, passing the messages that they generate during runtime to the check method of the entity. The entity would broadcast the passed message to its components and they would validate/invalidate the action. Return signals would go up from leaves to the root of the BT and, as a result of this, failures and warnings would be located and associated with one branch of the BT (depending on the decision nodes). Therefore, how these failures and warnings would be fixed or reported in design time would depend on how the tool works. Nevertheless, the easier way for solving a warning during runtime is to remove the whole branch, while failures must directly invalidate the association between the entity and the BT.

Once all this infrastructure is working, it is easy to use it as a sanity check for all the runtime retrieved BTs for our *query nodes*. In this way, we avoid blindly trying to run a BT that will fail later because the entity is unable to execute some of the primitive actions. The next section will describe a detailed example of the whole process.

5 Example

Let us imagine a shooter game in which a soldier watches over the approach roads of a bunker, patrolling the area and killing the enemies (players) without being seen whenever it is possible. The behaviour executed by this NPC can be the one shown in Fig. 4, which represents the BT corresponding to the goal Guard Area. This BT has two branches, one to patrol and another to kill the enemy. As there are several ways to kill an enemy, and the chosen way will depend on the virtual environment, the NPC type and its parameters, the Guard Area Behaviour BT has a *query node* to choose the attack behaviour.

On the other hand, Fig. 5 shows the Patrol Soldier entity type made up of components with its default attributes. During the game, an entity of the Patrol Soldier type, among others, will carry out the Guard Area BT (Fig. 4); so, as we will see, the attack behaviour will be chosen accordingly.

The execution context of the Guard Area BT is composed of three variables: ?this, ?world and ?target. ?this denotes the NPC executing the BT and its attributes describe its properties. ?world denotes the virtual environment in which the game takes place. ?target is an input variable for the Attack behaviour and denotes the entity targeted for this behaviour.

To execute this behaviour, the node Guard Area is executed. Being a *dynaymic priority list*, it will try to execute the first of its children and, if it is not possible, it will pass to the next one. The first child of the Guard Area node is a sequence, so it tries to execute all of its children, one after the other, beginning with Is Entity

Seen. Now suppose that Is Entity Seen fails (there are no visible entities of type PLAYER). This makes the sequence to fail and the next behaviour in the priority list, Patrol, is then executed. As the Guard Area behaviour is a *dynamic* priority list, it will keep trying to execute the first child (the sequence) in the subsequent cycles.

While executing Patrol, let us suppose that a PLAYER is seen by the NPC. The Patrol behaviour is interrupted to launch again the sequence. It tries to execute again the behaviour Is Entity Seen, succeeding this time. The entity detected by Is Entity Seen is stored in the attribute ?this.target, and the next behaviour in the sequence is executed. The next behaviour is a query, so it has to be solved to a BT before it can be executed.

The attributes for this query are:

1. *Header*: The name of the query behaviour (Attack (?target)") and a description ("Query behaviour Attack").
2. *Domain*: The retrieved behaviours should belong to the class Attack or to any of its children.
3. *Parameters*: The retrieved behaviour has to have an input parameter, target, which should be applicable to an entity of class PLAYER.
4. *Descriptors*: This attribute lists the game state descriptors that are considered relevant for the query:
   - ?target.distance: The distance to the target entity.
   - ?this.personality: The personality attribute of the entity executing the behaviour.
   - ?this.aggressive and ?this.defensive: The aggressiveness and defensiveness levels of the entity.
   - ?this.health: The health of the entity.
   - ?this.underAttack: Measures the attack received by the entity, being LOW when it is not being attacked and HIGH when being attacked by several entities in the close range.
   - ?world.time: The current time in the simulation environment (NIGHT or DAY).
5. *Weights*: The similarity section of the query refers to the importance of the descriptors for this query. The distance, personality of the entity and the fact of being attacked (the underAttack condition) are very important. The health and the current time are important. The aggressiveness and defensiveness are taken into account, but they are not as important as the rest.
6. *Requery*: The query has to be repeated when there is a significative change in the time, health or underAttack descriptors.
7. *Cases retrieved*: The query will retrieve all the cases in the case base.

To retrieve a BT, we have to compare the query with all the cases in the case base. First, we filter the case base using the *Goal* attribute, keeping only the cases that belong to the *Goal* class or any of its subclasses. In the next step, we also take

Table 2 Game state and similarity values

(a) Game state

Descriptor	Game state
target.distance	HIGH
this.personality	STEALTHY
this.aggressive	HIGH
this.defensive	MEDIUM
this.health	MEDIUM
this.underAttack	LOW
world.time	NIGHT

(b) Similarity

Case	Similarity
C_1	0,90
C_2	**1,00**
C_3	0,42
C_4	0,60
C_5	0,39

away all the cases with parameters that are not compatible with the ones in the query. Finally, the values of the descriptors of the cases are compared with the values of the relevant descriptors of the game state. Using the weights, a similarity value is obtained for each case.

For instance, let us imagine a night situation in which the NPC is close enough to a player to detect it but not so close, thus the player has not seen and attacked it yet. After filtering the case base, the query process retrieves the set of cases shown in Table 1 (the ones for the *Attack* goal). Then, every case has to be compared with the query. The values of the relevant descriptors of the query are retrieved from the game state and compared to the corresponding descriptors in the cases. Table 2a, b shows the values of the relevant descriptors for our example query and the results of calculations of the similarity values for each case and the query. As it is shown in the table, stealth behaviours are predominant over the rest because of the night situation, the personality of the NPC and due to the fact that the NPC is not under attack. `Long Range Stealth Attack` has better score than `Hand To Hand Stealth Attack` just because of the distance between the NPC and the player.

Once the set of cases has been retrieved and ordered by its similarity, the query process must return the most similar behaviour to the query but, at the same time, the NPC must be able to carry out this behaviour. There should be taken into account that the behaviours stored in the case base may not be suitable for every entity. Different entity types will have different abilities, and even entities of the same type could have different parameter values (e.g. strength).

Here is where the *reflective components*, described in Sect. 4, become useful. When the first part of the query process ends up with a list of BTs ordered by its similarity with the query, the query process iterates over them looking for the first one that may be executed by the actual entity. The query process will finally return the behaviour most similar to the query that can be carried out by the NPC.

In our example, the query process has to check which of the retrieved BTs can be executed by the `Patrol Soldier` entity, whose components are listed in Fig. 5. We will reduce our explanation just to the two most similar BTs retrieved from the query, `Long Range Stealth Attack` and `Hand To Hand Stealth Attack`, which appear in Fig. 6a, b. These BTs need special skills to be carried out, so the query process must assure, by means of our *reflecting components*, the NPC will be able to execute these retrieved BTs. When validating `Long Range`

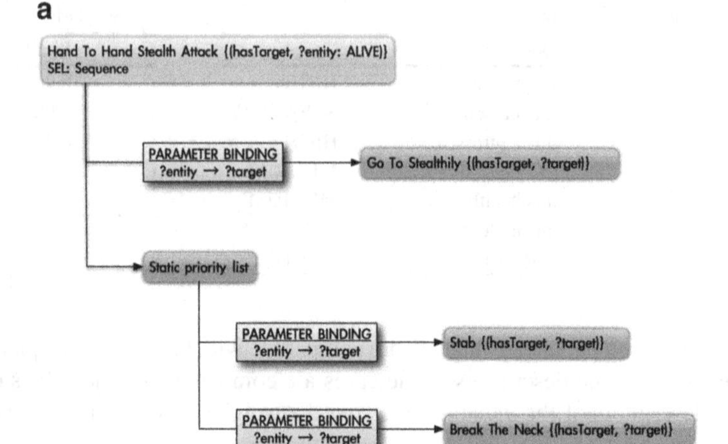

Hand To Hand Stealth Attack

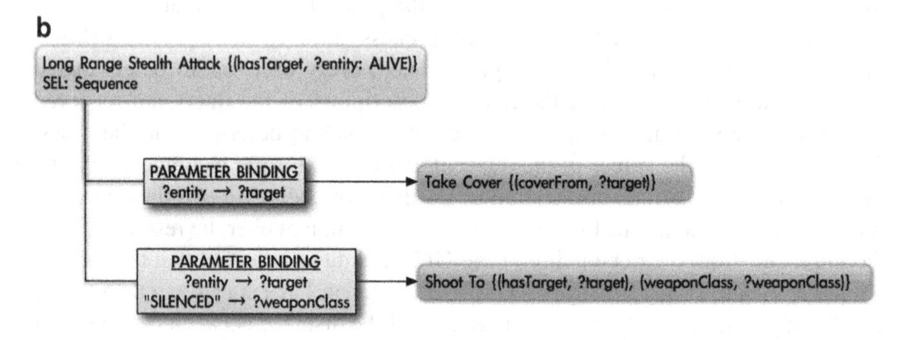

Long Range Stealth Attack

Fig. 6 BTs of the example

`Stealth Attack` before its execution, the system detects that the `Shoot To` action cannot be carried out by the NPC. Although a `Patrol Soldier` has a `ShootTo` component that allows long-range attacks with firearms and a `Patrol Soldier` has a rifle and a gun, it does not have a silencer and thus the action will not be successful, and consequently this BT is rejected.

Then is the turn of the `Hand To Hand Stealth Attack` BT. In the same way, when validating `Hand To Hand Stealth Attack` before its execution, the system detects that the `Stab` action cannot be carried out by the NPC. In this case, the failure returned by the `Stab` action is because the NPC does not have a sharp arm like a knife. The failure is propagated to the sequence node; however, in this case, the failure is not propagated further on because the static priority list node has another valid choice to execute: the `Break The Neck` action.

Consequently, the `Hand To Hand Stealth Attack` BT is the behaviour returned due to the fact that it is the most similar behaviour to the query that *can* be executed by the NPC. Once the retrieval process has ended, the execution of the original BT continues in the query node of the original BT and it executes the recovered `Hand To Hand Stealth Attack` BT in a transparent way.

6 Conclusions

BTs are a great tool for design AI game behaviour because they have an easy graphical representation and promote reuse of complete or partial BTs based on their hierarchical nature. Unfortunately, they intrinsically include some programming concepts that provide them with the expressive power of a general purpose programming language, making them difficult to understand for non-technical designers. As a consequence, BTs are mainly used by *programmers*, who *draw* the behaviours instead of just writing down in some concrete programming language. Designers are, usually, in charge of very high-level BTs, with just a few nodes that are easy to create and debug. They are built putting together more complex BTs created by programmers; so designers must have an easy access to the *library of BTs* where all the BTs created for the game team are stored.

This can be, in fact, quite complex. At the end of the game developing cycle, the team can have produced a quite high amount of BTs, where designers (and also programmers) must dive into in order to look for concrete behaviours while creating new ones. Some kind of *automatic search* is welcome in the BT library to alleviate the time spent while looking for BTs. In this chapter, we have presented a tool for BT design that includes such a feature, using CBR techniques to retrieve the more adecuate BTs [4].

On the other hand, as was stated in this chapter, during game production AI designers create *BTs* mixing the basic behaviours with aggregation in BTs. *At the same time, developers* create new basic behaviours depending on the ongoing necessities (the `Stealth attack` of our example would be one of them). As a result, designers will have more basic behaviours to play with at the end of the production, and the last created BTs will be richer than the first ones.

The ad hoc solution for this consistent problem is to revise the older BTs for detecting whether they could be improved using the more recent basic behaviours created by the development team. Unfortunately, this revision effort needs a lot of time and should be performed during all the game production timeline.

Using our *query nodes* [5], on the contrary, old BTs are automatically benefited from new behaviours if they are correctly stored and annotated in the case base. The example has shown that, when using our technique in the `Attack` node, no revision is needed if a new `Stealth Attack` behaviour is developed.

The main advantage of our proposal is that the number of *basic behaviours* can grow throughout the game development and, even so, be quite sure that they will be used in older complex behaviours. Having this confidence when using static BTs requires a manual revision of the previous developed BTs, something only affordable if the number of added behaviours is kept low. Consequently, our proposal provides a better scalability for the *growth of basic behaviours*.

As a welcomed secondary effect, and due to the fact that the query nodes take into account all the basic behaviours in the case base, BTs using them could provide richer behaviours with no design effort. The manual alternative would require the substitution of our *query node* with a priority list (as in the example) with all the available basic behaviours. Again, this becomes impractical, demonstrating that *query nodes* provide a better scalability also *in the number of basic behaviours considered at run-time*.

Unfortunately, all these advantages do not come for free. The cost for this saving is, obviously, categorizing each new basic behaviour for the *query node* to recover it in the correct moments. Behaviour and entities ontologies (the vocabulary for describing our cases) must also be created, although they could be reused between projects (after all, reuse is one of the goals of ontologies).

At run-time, our *query node* will spend more time *the first time* for extracting the appropriate basic behaviour if comparing with a priority list. But, due to the *re-query* attribute in the query node, we avoid spending time every AI cycle to change the first election, something that priority lists do not do. On the other hand, debug behaviours using our *query nodes* will be a bit more complex due to the new uncertainty ingredient added to the behaviour selection. This problem can, in fact, be seen as an advantage, because some *emergent behaviour* usually is considered to provide game variability.

Acknowledgements Supported by the Spanish Ministry of Science and Education (TIN2009-13692-C03-03) Funded by Complutense University of Madrid.

References

1. Buchanan, W.: Game Programming Gems 5, chap. A Generic Component Library. Charles River Media, Boston, MA (2005)
2. Crytek: CryENGINE 3 SDK, Sandbox Editor (2010). Http://mycryengine.com
3. Esmurdoc, C.: Head games: Double fine's psychonautic break. Game Developer Magazine **12**(7), 30–38 (2005)
4. Flórez-Puga, G., Díaz-Agudo, B., González-Calero, P.A.: Experience-Based Design of Behaviors in Videogames. In: K.D. Althoff, R. Bergmann, M. Minor, A. Hanft (eds.) ECCBR, *Lecture Notes in Computer Science*, vol. 5239, pp. 180–194. Springer, Berlin (2008)
5. Flórez-Puga, G., Gómez-Martín, M.A., Díaz-Agudo, B., González-Calero, P.A.: Dynamic Expansion of Behaviour Trees. In: C. Darken, M. Mateas (eds.) AIIDE. The AAAI Press, Menlo Park, California (2008)
6. Games, E.: Unreal Development Kit (2009). Http://www.udk.com/
7. Garcés, S.: AI Game Programming Wisdom III, chap. Flexible Object-Composition Architecture. Charles River Media, Boston, MA (2006)
8. Hocking, C.: Ubisoft Montreal's Far Cry 2 Postmortem. Game Developer Magazine, pp. 30–38 (2009)
9. Ierusalimschy, R.: Programming in Lua, second edn. Lua.org (2006)
10. Isla, D.: Handling Complexity in the Halo 2 AI. In: Game Developers Conference (2005)
11. Lakos, J.: Large Scale C++ Software Design. Addison Wesley, Reading (1996)
12. Llansó, D., Gómez-Martín, M.A., González-Calero, P.A.: Self-validated behaviour trees through reflective components. In: Proceedings, The Fifth AAAI Artificial Intelligence and

Interactive Digital Entertainment Conference, pp. 76–81. AAAI Press (2009). URL http://www.aaai.org/Conferences/AIIDE/aiide09.php
13. Millington, I., Funge, J.: Artificial Intelligence for Games, second edn. Morgan Kaufmann, San Francisco, California (2009)
14. Rabin, S. (ed.): AI Game Programming Wisdom 3. Charles River Media, Boston, MA (2006)
15. Rabin, S. (ed.): AI Game Programming Wisdom 4. Charles River Media, Boston, MA (2008)
16. Rene, B.: Game Programming Gems 5, chap. Component Based Object Management. Charles River Media, Boston, MA (2005)
17. Szyperski, C.: Component Software: Beyond Object-Oriented Programming. Addison-Wesley Professional, Boston, MA (1997)
18. Tai, K.C.: The tree-to-tree correction problem. J. ACM **26**(3), 422–433 (1979). DOI http://doi.acm.org/10.1145/322139.322143
19. Wang, J.T.L., Shapiro, B.A., Shasha, D., Zhang, K., Currey, K.M.: An algorithm for finding the largest approximately common substructures of two trees. IEEE Transactions on Pattern Analysis and Machine Intelligence **20**, 889–895 (1998)
20. West, M.: Evolve your hierarchy. Game Developer **13**(3), 51–54 (2006)
21. Zhang, K., Shasha, D.: Simple fast algorithms for the editing distance between trees and related problems. SIAM J. Comput. **18**(6), 1245–1262 (1989). DOI http://dx.doi.org/10.1137/0218082
22. Isla, D.: Halo 3 - building a better battle. In: Game Developers Conference (2008)
23. Krajewski, J.: Creating all humans: A data-driven AI framework for open game worlds. Gamasutra (2009)
24. Atkin, M.S., King, G.W., Westbrook, D.L., Heeringa, B., Cohen, P.R.: Hierarchical agent control: a framework for defining agent behavior. In: Agents, pp. 425–432 (2001)

Game AI for Domination Games

Chad Hogg, Stephen Lee-Urban, Héctor Muñoz-Avila, Bryan Auslander, and Megan Smith

Abstract In this chapter, we present an overview of several techniques we have studied over the years to build game AI for domination games. Domination is a game style in which teams compete for control of map locations and has been very popular over the years. Due to the rules of the games, good performance is mostly dependent on overall strategy rather than the skill of individual team members. Hence, this makes domination games an ideal testbed to study game AI.

1 Introduction

Domination is a game style in which teams of players compete to control certain locations on a map called *domination points* within a real-time environment. Specifically, a domination point is controlled by the team whose player last stepped on it. Each second the teams earn points for each of the domination points that they currently control and have controlled for all of some preceding time window. In addition to moving around on the map, players are able to engage in combat with players from the opposite team when they are nearby. A player who is killed in combat respawns at a point randomly selected from a set of pre-determined *spawn points* on the map. In the meantime, the player who killed them may be able to take advantage of this to gain control of a domination point that the killed player had been defending.

Domination games have been used, either exclusively or as an option, in a variety of game genres, including first-person shooters (e.g., Half-Life®, Call of Duty®), role-playing games (e.g., World of Warcraft®), and third-person shooters (e.g., Gears of War 2®). In addition, many other games that do not perfectly fit the domination model have similar characteristics. For example, Counter-Strike®, which is among the most popular multiplayer games ever released, also consists of two teams competing in a real-time environment in which strategic control of certain locations (such as bomb sites and hostage drop-offs) is vital to success.

C. Hogg (✉)
Department of Computer Science & Engineering, Lehigh University, Bethlehem, PA 18015, USA
e-mail: cmh204@lehigh.edu

P.A. González-Calero and M.A. Gómez-Martín (eds.), *Artificial Intelligence for Computer Games*, DOI 10.1007/978-1-4419-8188-2_4,

Although an individual player who is highly skilled in combat gives his team a definite advantage, this is much less true than in other game types where, for example, points are awarded for each kill. Instead, team-based strategy is of high importance in these games. Although no player has control over the actions of his teammates, many of these games have built-in communication systems to allow players to devise and execute specific strategies with their teammates. We suspect that the team-oriented nature of these games is a primary reason for their enduring success, and it also makes them an excellent testbed for AI development.

In addition, domination games can generally be classified as having the following properties: Domination games are *non-deterministic*; success in combat requires both skill and luck, and it is not possible to predict whether or not a player will successfully reach his objective. Domination games are also *adversarial*; two or more teams compete to control the domination points. Finally, domination games are *imperfect information* games; a team only knows the locations of those opponent players that are within the range of view of one of the team's own players. These conditions make domination games a good testbed for evaluating algorithms that integrate planning and execution.

We refer to individual players who are not human-controlled as *bots*. The purpose of our research is not to improve the combat performance of individual bots, and so we use the same Finite State Machine-based bot logic for all of our computer-controlled players. Our interest is in the overarching strategies that teams of bots pursue.

Over the years, we have devised several methods that integrate planning and execution for selecting a team's strategy in domination games. Table 1 shows a summary of the three algorithms that we will discuss in this chapter: HTNBOTS, RETALIATE, and CBRETALITE. HTNBOTS uses hierarchical task network (HTN) representation techniques to generate new plans [3]. It monitors the current situation in the game; when the circumstances change, it generates new plans on the fly. RETALIATE uses reinforcement learning (RL) techniques; it uses a Q-learning algorithm to find policies that represent competent ways to play the game [16]. The third system is CBRETALITE [1]. CBRETALITE is built on top of RETALIATE; it stores and retrieves a library of policies, which are reused by the RL algorithm from RETALIATE.

We have conducted a study that compares these three approaches. In this chapter, we report on the architectures of these three systems and comparisons among the knowledge requirements and performance results of the three systems.

Table 1 Three systems for playing domination games

Game AI	Description
HTNBOTS	Replanning algorithm. Uses HTN planning techniques to generate plans on-the-fly
RETALIATE	Generates policies that adapt to the opponent using RL techniques
CBRETALITE	Stores and reuses policies generated by RETALIATE

2 DOM: A Generic Domination Game Environment

Our initial experiments with each of our three systems used the commercially available game Unreal Tournament® as the simulation environment. While this simulator provides a useful API for controlling teams of bots, we found that a number of factors made it difficult to perform large-scale experiments and extract useful data from them.

Instead, we built a game environment, called DOM, which captures the essence of domination games [1]. The basic rules in DOM are the following: Each time a bot on team t passes over a domination point, and that point will belong to t. Team t receives one point for every five game ticks that it owns a domination point. Teams compete to be the first to earn a predefined number of points. No awards are given for killing an opponent team's bot, which *respawns* immediately in a location selected randomly from a set of map locations, and then continues to play. A location is captured by a team whenever one of its bots moves on top of the location, and within the next five game ticks no bot from another team moves on top of that location.

The total number of possible states in the game is at least $O(2 \times 10^{34})$, assuming a standard map of 70×70 cells, four domination locations, and three bots per team [2]. It would be infeasible for our agents to reason in such a complicated world; so we have used an abstraction of states and actions that was first described in the work on RETALIATE [16]. In the abstracted description of the world, the current state consists only of the ownership of each domination point; each point is either owned by one of the teams or is unowned at any point in time. Thus, for an environment with d domination points and t teams, the total number of possible states is d^{t+1}. We also abstract away the possible actions of the bots. In this abstraction, each action states only to which domination location each of the bots on a team should go. Thus, for an environment with d domination points and b bots per team, the number of actions available to a team is b^d. The details of how a bot moves from one location to another are strictly determined by a shortest-path algorithm.

3 The Game AI Systems

We briefly summarize each of the three algorithms we investigated that integrate planning and execution for playing DOM. Each of the algorithms focus on controlling which domination locations team-member bots are sent to. Consequently, the behavior of the individual bots can be pre-determined by a standard FSM. Our algorithms do not make a priori assumptions about what that behavior is, which allows bots to be used as plug-ins. In principle, this allows the design decisions for the team AI to be made independently of the design decisions relating to the control of individual bot behavior. Similarly, using bots as plug-ins, the game developer can swap different bot types in and out of the game, and even use bots developed for single-player non-team modes in multi-player games. For further details of these algorithms, please see the references.

3.1 HTNBOTS

HTNBOTS is a dynamic replanning algorithm that uses HTN planning techniques to generate plans [3, 8]. HTN planning proceeds by decomposing high-level tasks such as *win domination game* into simpler tasks such as *send bot* $b1$ *to location* $L1$. There are two kinds of tasks: compound and primitive. Compound tasks, such as *win domination game*, can be further decomposed into subtasks whereas primitive tasks cannot. The primitive tasks denote concrete actions, such as *send bot* $b1$ *to location* $L1$. Each level in an HTN adds detail on how to achieve the high-level tasks. The sequencing of the leaves in a fully expanded HTN yields the plan for achieving the high-level tasks.

3.1.1 Planning Knowledge in HTNBOTS

HTN planners require that the planning knowledge be provided in the form of methods and operators. A method encodes how to achieve a compound task and consists of three elements:

- *Head*: The task being achieved, called the *head* of the method.
- *Preconditions*: The set of preconditions indicating the conditions that must be fulfilled for the method to be applicable.
- *Subtasks*: The subtasks needed to achieve the head.

Table 2 shows an example of a method (*?<string>* indicates that *<string>* is a variable). The task that this method achieves is that team T gains control of locations $?L1$ and $?L2$. This method is applicable when the variables $?L1$ and $?L2$ refer to domination locations and the variables $?bot1$, $?bot2$, and $?bot3$ refer to distinct bots on team T. The method accomplishes its head by ordering one of the bots to go to location $?L1$, another to go to location $?L2$, and the third to patrol between those two locations.

Table 2 Example method and operator in HTNBOTS

Method	Operator
Head: Control2Locations(T, $?L1$, $?L2$)	Head: sendBot(b, $?LD$)
Preconditions:	Preconditions:
domLocation($?L1$)	botLocation(b, $?LC$)
domLocation($?L2$)	Effects:
teamMember($?bot1$, T)	$\neg$ botLocation($?b$, $?LC$)
teamMember($?bot2$, T)	botLocation($?b$, $?LD$)
teamMember($?bot3$, T)	
different($?bot1$, $?bot2$, $?bot3$)	
Subtasks:	
sendBot($?bot1$, $?L1$)	
sendBot($?bot2$, $?L2$)	
patrol($?bot3$, $?L1$, $?L2$)	

Operators define valid actions in the domain. An operator consists of:

- *Head*: The primitive task that the operator accomplishes.
- *Preconditions*: The conditions that must be true for the operator to be applicable.
- *Effects*: How the current situation changes as a result of applying the operator.

Table 2 also shows an example of an operator. This operator sends a bot b to a location $?LD$. The operator is applicable when bot b is at location $?LC$. After successful execution of the operator, bot b will no longer be in location $?LC$, and will instead be in location $?LD$. The preconditions and effects of operators allow the planner to construct a plan that will achieve the primitive task *if* the action is executed successfully. Because these plans are executed in a nondeterministic environment, this is not always the case.

3.1.2 HTN Planning in HTNBOTS

The following are the steps performed by HTNBOTS to decompose a compound task, t:

1. $M \leftarrow$ select all methods whose head matches t
2. $m \leftarrow$ select a method from M that is applicable
3. decompose t with the subtasks of m

For selecting an applicable method, m, HTNBOTS checks whether the preconditions are valid in the current state of the game world. This is accomplished through a communication protocol between HTNBOTS and the game engine. For example, for the method shown in Table 2, each of its six preconditions must be fulfilled in the game world, which will also result in the variables being instantiated to concrete objects in the game world. If the method is used, then it will decompose the task Control2Locations (T, $?L1$, $?L2$) (with proper instantiation of the variables) into the three subtasks indicated in the method.

The three steps above are repeated recursively for each compound task in the subtasks of m until a primitive task is reached. For achieving a primitive task, HTNBOTS performs the following steps:

1. $O \leftarrow$ select all operators whose head matches t
2. $o \leftarrow$ select an operator from O that is applicable
3. Execute action for operator o

Once again the communication protocol is used to determine whether an operator is applicable by checking whether the operator's preconditions are valid in the current state of the game world. The communication protocol is also used to execute the action indicated by the operator. An action is a ground instance of an operator. That is, the operator's variables are instantiated with objects (the particular objects are identified by the game engine when determining whether the operator is applicable). The game engine has code for executing actions. For example, if the action is sendBot(*b*33, *L*77), then the game engine will compute a path from the current location of bot *b*33 to location *L*77. The code for executing the action will also determine what to do in in-game situations such as encountering an opponent.

3.1.3 Plan Execution

In HTNBOTS each action indicates a concrete activity to be executed by one bot. As a result, HTNBOTS can execute actions in parallel if these are performed by different bots. Once the plan is generated, HTNBOTS will start executing each action in the order indicated by the plan. The following steps are performed for each action *a* in the plan:

1. Check whether the bot *b* assigned for performing *a* is performing another action; if not then execute *a*.
2. If *b* is performing another action, then wait until *b* is done, and then execute *a*.

These steps ensure consistency in the execution of the plan. On the other hand, they might unnecessarily delay the executions of other actions (e.g., those actions to be performed by other bots that occur later in the plan). More sophisticated execution control could be implemented (e.g., if a latter action is not dependent on a currently delayed action, it could be executed).

When a plan is executed, HTNBOTS keeps constant track of the preconditions of the method decomposing the top-level task (i.e., to win a domination game). If a percentage of these preconditions that are no longer valid is greater than a predefined threshold, a new plan is generated and executed. The rationale is that we only want to change the plan if enough conditions in the game have changed making it necessary to adapt to these changes. Typically methods decomposing the top-level tasks have preconditions about ownership of the domination locations, and the method indicates strategies for dealing with those situations. When domination location ownership changes substantially, it makes sense to immediately generate a new plan to adapt to the new situation. HTN plan generation in HTNBOTS is extremely fast making this process seamlessly.

3.2 RETALIATE

We used RL to create RETALIATE which uses Q-learning [15] to acquire winning strategies for games in DOM. Unlike some other forms of learning, RL does not require annotated training examples to learn, nor does RL need an expert to provide feedback to the learner. In RL, interaction with the world is the only way the agent gains information: the agent (1) senses the state of the environment, (2) chooses which action to take, (3) performs the action, and (4) receives a (scalar) reward or punishment. Under the RL approach, time is spent crafting the representation of the game state, called the "problem model" – that is, how the various complexities of complete game states are abstracted into a simpler form that RL can use. This is typically significantly easier than manually designing and implementing strategies in complete symbolic representations, such as HTNBOTS . The problem model used by RETALIATE is presented in Sect. 3.2.2.

In this section, we first briefly describe RL in general and Q-learning in particular. Next, we present the way in which we modeled the states and actions of DOM to increase the efficiency of our application of the Q-learning process. This section is concluded with the RETALIATE algorithm.

3.2.1 Reinforcement Learning

RL is a form of machine learning where an agent or team of agents learns a policy – what action to select in every perceived world state – in a potentially stochastic environment. The goal in RL is to arrive at an optimal policy which is the one that maximizes the rewards received, through a process of trial and error. For an overview of RL in general, see [15].

The purpose of RL algorithms is to find a policy π that maximizes the sum of the returned rewards. A policy π is a mapping from states to actions indicating for each state s, the action $\pi(s)$ that should be chosen. This mapping is calculated using the rewards received from previous action selections in states already visited. Each state-action value is a representation of the expected future rewards of taking s in a, and assumes that policy π is used for all subsequent action selections. Rewards are obtained from the environment as a result of the agent's actions and are measured as $U(s') - U(s)$, the difference between the utilities of the current state s and the next state s' that will be reached after executing an action. In Sect. 3.2.2, we present our definition of the utility and reward functions for use of the RETALIATE algorithm in DOM.

In RL, including Q-learning, the choice of which action to take in state s, that is $\pi(s)$, involves the use of estimates of the expected value of taking each action in every possible game state. These estimates are derived from the rewards received after taking a selected action in a given state. There exist multiple ways for keeping track of the estimates, and in Q-learning the most straightforward approach is to maintain a "Q-table" that associates with each $(state, action)$ pair the estimated value $Q(s,a)$ of the pair, called the "Q-value". The value of the reward, R, is used to perform an update on the Q-table entry $Q(s,a)$ for the previous state s in which the last action a was ordered. The Q-table approach is only feasible when the number of states and actions in the problem model is limited, not only because the table size can become very large otherwise (the size of the table is the number of states multiplied by the number of actions), but also because the amount of learning cycles required to arrive at an accurate estimate of the Q-value grows with the number of state-action pairs.

The update on the Q-table entry $Q(s,a)$ for the previous state s in which the last action a was executed is computed according to the following formula, which is standard for updating the entries in a Q-table in temporal difference learning:

$$Q(s,a) = Q(s,a) + \alpha\{R + \gamma * \underset{a'}{\operatorname{argmax}}[Q(s',a') - Q(s,a)]\} \quad (1)$$

In this computation, the Q-value in the Q-table for the action a that was just taken in state s, $Q(s,a)$ is updated. The function argmax returns the value from the Q-table of the best team action that can be performed in the new state, s', which is simply the highest value associated with s' in the table for any a'. The value of γ, which is called the discount-rate parameter, adjusts the importance of future rewards in making current decisions.

In order to safeguard against creating Q-values (and therefore policies) that are stuck in a local optimum, action selection is often performed using an "ε-greedy strategy"; rather than always executing the action of highest estimated value in a given state, when the system is in state s, with a probability of $1-\varepsilon$, it selects the action a with the highest $Q(s,a)$, and with a probability ε it selects a random action. The policy π used by a Q-learning agent in this case is therefore the combination of ε-greedy action selection with a Q-table.

3.2.2 Problem Model: Definition of States and Actions

When deciding upon the problem model to use in RL, one must consider the essential features of the problem being addressed. For example, while the amount of ammunition remaining is important for an individual team member, the overall team's strategy might safely ignore this detail. A problem model that takes into consideration too many features of the game state can lead to a learning problem that is very difficult, or impossible, for the system to solve in a reasonable amount of time. Similarly, an overly simplified problem model leads to a system that does not play very well, or one that has very limited capabilities. The trick is to model the problem in such a way that learning can happen quickly, while simultaneously being rich enough to support a range of interesting behaviors.

In RETALIATE, game states are represented in the problem model as a tuple indicating the owner O_i of the domination location i. For instance, if there are three domination locations, the state (E,F,F) describes the state where the first domination location is owned by the enemy and the other two domination locations are owned by our friendly team. Neutral ownership of a domination location is also considered and is represented by an N in the relevant location in the tuple. For three domination locations and two teams, there are 27 unique states of the game, taking into account that domination locations are initially not owned by either team.

The addition of other parameters was considered to increase the information contained in each state. The additional information slowed the RETALIATE learning process considerably, reduced the effectiveness of the RL team, and ultimately was not worth the additional computational cost. In contrast, not only did the simpler definition greatly reduce size of the state space, leading to more rapid learning, but, as our result show, it also contained sufficient information to develop a winning policy. The separation of parameters – those used to define team tactics versus those used for individual behavior – is one of the central qualities of RETALIATE.

In RETALIATE, states are associated with a set of *team actions*. A team action is defined as a tuple indicating the individual action A_i that bot i takes – for a team of three bots, a team action tuple consists of three *individual actions*. An individual action specifies to which domination location a bot should move. For example, in the team action $(Loc1, Loc2, Loc3)$, the three individual actions send $bot1$ to domination location 1, $bot2$ to domination location 2, and $bot3$ to domination location 3, whereas in $(Loc1, Loc1, Loc1)$, the individual actions send all three bots to domination location 1. If a bot is already in a location that it is told to move to, the action is interpreted as instructing the bot to stay where it is. Individual bot actions are executed in parallel and, for a game with three domination locations and three bots, there are 27 unique team actions because each bot can be sent to three different locations. The Q-table therefore contains $27 \times 27 = 729$ Q-value entries.

Despite the simplicity in the representation of our problem model, it not only proves effective but it actually mimics how human teams play domination games. The most common error of novice players in this kind of game is to fight opponents in locations other than the domination ones; these fights should be avoided because they generally do not contribute to victories in these kinds of games. Part of the reason for that is that if a player is killed away from a domination location, it will not have a direct effect on ownership and hence will not have an effect on the score. Consequently, it is common for human teams to focus on coordinating to which domination points each team member should go, and this is precisely the kind of behavior that our problem model represents.

3.2.3 The RETALIATE Algorithm

Algorithm 1 presents pseudocode for the RETALIATE online learning algorithm. RETALIATE is designed to run across multiple game instances so that the policy, and therefore the RETALIATE -controlled team, can adapt continuously to changes in the environment while keeping track of what was learned in previous games.

Algorithm 1 RETALIATE(Q_t)

1: **Input**: Q-Table Q_t
2: **Output**: updated Q-table
3: ε is 0.1, α is 0.2, γ is 1.0, and state s_{prev} is maintained internally
4: **if** $\text{rand}(0,1) > \varepsilon$ **then** {epsilon greedy selection}
5:     action $a \leftarrow$ applicable action with max value in Q-table
6: **else**
7:     action $a \leftarrow$ random applicable action from Q-table
8: **end if**
9: state $s_{\text{now}} \leftarrow \text{Execute}(a)$
10: reward $R \leftarrow U(s_{\text{now}}) - U(s_{\text{prev}})$
11: $Q_t(s_{\text{prev}}, a) \leftarrow Q_t(s_{\text{prev}}, a) + \alpha(R + \gamma max_{a'} Q_t(s_{\text{now}}, a') - Q_t(s_{\text{prev}}, a))$
12: $s_{\text{prev}} \leftarrow s_{\text{now}}$
13: return Q_t

These changes can include changes in our own players (e.g., different type of bot), changes in the opponent team (e.g., changes of tactics), and changes in the game world (e.g., a new map).

RETALIATE is controlled by three Q-learning parameters: the "epsilon-greedy" parameter ε, which controls the tradeoff between exploration and exploitation by setting the rate at which the algorithm selects a random action rather than the one that is expected to perform best, the "step-size" parameter α, which influences the rate of learning, and the "step-size" parameter γ, which determines the present value of future rewards. For our empirical evaluations, we found that setting ε to 0.1 and α to 0.2 works well. RETALIATE diverges from the traditional discounting of rewards by setting γ equal to one so that possible future rewards were as important as in selecting the current action as immediate rewards. Initially, we set $\gamma < 1$ to place an emphasis on immediate rewards but found that the rate of adaptation of RETALIATE was slower than when γ was set to one.

RETALIATE starts by either initializing all entries in the Q-table with a default value, which was 0.5 in our case study, or restoring the Q-table from the previous game. The game is then started, and the game state representation s_{prev} is initialized to each domination location having neutral ownership (N,N,N).

The following computations are iterated through until the current game is over. First, the next team action to execute, a, is selected using the epsilon-greedy parameter; this means that a random team action is chosen with probability ε, or the team action with the maximum value in the Q-table for state s is selected with probability $1-\varepsilon$. By stochastically selecting actions, we ensure that there is a chance of trying new actions, or trying actions whose values are less than the current maximum in the Q-table. This is important to ensure that RL experiments with a wide range of behaviors before deciding which is optimal.

The selected action a is then executed, and the resulting state, s_{now}, is observed. Each bot can either succeed in accomplishing its individual action or fail (e.g., the bot is killed before it could reach its destination). Either way, executing a team action takes only a few seconds because the individual actions are executed in parallel. Updates to the Q-table occur when either the individual actions have completed (whether successfully or unsuccessfully), or domination location ownership changes because of the actions of the opposing team.

Next, the reward R for taking a in s_{prev} is computed as the difference between the utilities in the new state s_{now} and the previous state s_{prev}. The reward function, which determines the scale of a reward, is computed as $R = U(s_{\text{now}}) - U(s_{\text{prev}})$. Specifically, the utility of a state s is defined by the function $U(s) = F(s) - E(s)$, where $F(s)$ is the number of friendly domination locations and $E(s)$ is the number of enemy-controlled domination locations. For example, relative to team A, a state in which team A owns two domination locations and team B owns one domination location has a higher utility than a state in which team A owns only one domination location and team B owns two.

Finally, the Q-value $Q_t(s_{\text{prev}}, a)$ for taking action a in state s_{prev} is used in the standard Q-table update function presented in (1). Having completed the current update, the new state s_{now} is backed up in variable s_{prev} for the next update, and the modified Q-table is returned.

3.3 CBRETALITE

One of the limitations of RL agents in general and RETALIATE in particular is that the process of converging to an optimal policy may be slow. Worse, when the situation changes in a way that is not reflected directly in the states observed by the agent, a policy that was previously optimal may no longer be a good choice, and the slow process of finding an optimal policy for the new problem must begin. This is a result of the trial-and-error process by which RL agents incrementally update their policies based on experience. It is possible to fine-tune the parameters of the Q-learning algorithm to adapt very quickly to changing conditions, but this has its own tradeoffs.

Instead, we developed a new system, CBRETALITE, that applies case-based reasoning (CBR) techniques to RETALIATE. The CBRETALITE system stores a library of cases, each of which contains a winning policy and the conditions under which that policy was learned. When the current policy is highly ineffective, the system searches for a case that matches the current situation and begins using the policy from that case. As a result, it is able to quickly change its strategy to counter the different strategies of a dynamic opponent.

3.3.1 Case-Based Reasoning

CBR is a general problem-solving strategy in which new scenarios are compared with problems that were previously solved, and a successful solution to a previous problem that is similar to the current scenario is adapted to solve the current problem. The knowledge artifacts that store information about previous problem-solving episodes are called *cases*, and typically consist of two components: a representation of the problem that was solved and a representation of the successful solution to the problem. A widely used model of CBR includes four steps: *Retrieve*, *Reuse*, *Revise*, and *Retain*.

In the *Retrieve* step, the system searches for one or more cases in its case library that have a problem that is similar to the current problem. In the *Reuse* step, a solution to the current problem is produced, either as a direct copy of the solution from a retrieved case, or adapted to take into account the differences there may be between the problem in the case and the current case. In the *Revise* step, the system updates its knowledge base as a result of the success or failure of the attempt to solve the current problem with the solution generated in the *Reuse* step. In the *Retain* step, a new case is inserted into the case library. This consists of a problem that was recently solved as well as the solution to that problem. This solution may have been generated from scratch, provided by a tutor, or produced from an existing case. Advanced CBR-based system may also have a feature to manage the case library, such as by removing redundant cases.

3.3.2 CBR in CBRetaliate

In CBRetaliate, the solution part of each case is a Q-table, as learned by Retaliate. The problem representation consists of a set of features describing the current game situation, which are not part of the state representation used within the Q-learning algorithm.

When a DOM game begins, there is very limited information available, but CBRetaliate selects a case based on what it does know. The Q-table from that case is used to make decisions during the game and is updated using the Q-learning algorithm exactly as in Retaliate. After a certain time window, CBRetaliate determines whether or not it has recently been successful, based on the rate at which each team's score has been changing.

If it has been highly successful, a new case is created and added to the case library. The problem section of this new case consists of the values of the relevant features at this time. The solution section of this new case consists of the Q-table that the agent is currently using, which is likely to have been updated somewhat in the time since it was copied from an existing case.

If CBRetaliate has been very unsuccessful, it will abandon the current Q-table and instead search for a case in the case library that is similar to the current situation and begin using its Q-table. If there is no case sufficiently similar to the current situation, or if CBRetaliate is neither winning nor losing by a significant margin, then it will neither store nor retrieve a case, but will continue using and updating its current Q-table.

If CBRetaliate has a new case to store and finds that there is already a case in the library with a very similar problem, it will compare how successful the agent was when each of the two cases was created, and will retain only the one with the Q-table that gave the agent the most success.

3.3.3 Features and Similarity

The representation of the problem used in the case library consists of several features that, based on observation and trial-and-error, seemed likely to correlate highly with the effectiveness of different strategies. Specifically, CBRetaliate uses the following features:

- *Team Size*: The number of bots on each team.
- *Team Score*: The current score of each team.
- *Bot Distance*: The distance between each bot in the game and each domination location.
- *Ownership*: The fraction of time over a rolling time window in which each domination location was owned by each of the teams.

To compute the level of similarity between one problem and another, CBRetaliate uses a *local similarity metric* for each feature type and a *global similarity metric* that aggregates the values of each local similarity metric. Local similarities are valued between zero and one and are computed by matching sensory readings from a time window within the current game world with those stored in the case. The value of the aggregate is simply the sum of the local similarity for each feature, divided by the number of features.

The *Team Size* feature type records the number of bots on a team. Teams are assumed to be of equal size; however, this assumption could be dropped using a feature for each team. If x is the size of the team in the current game and y is the team size from a case, $Sim_{\text{Tsize}}(x,y)$ is equal to one when $x = y$ and zero otherwise.

The *Team Score* feature type records the score of each team. Hence, if x is the score of team A in the current game and y is the score of team B from a case, then the similarity is computed by $Sim_{\text{TScore}}(x,y) = 1 - (|x-y|/SCORE\_LIMIT)$. The constant *SCORE_LIMIT* is the score to which games are played. In our case-base, team A is always CBRETALITE and team B is the opponent.

The next feature type, *Bot Distance*, uses the Euclidian distance of each bot to each domination location to compute similarity. That is, each case contains, for each opponent bot b and for each domination location l, the absolute value of the Euclidian distance from b to l. Specifically, if x is the Euclidian distance of b to l in the current game and y the analogous distance from the case, then $Sim_{Dist}(x,y) = 1 - (|x-y|/MAX\_DIST)$. The constant *MAX_DIST* is the maximum Euclidian distance any two points can be in a map. With an opposing teams of size 3 and a map with 3 domination locations, there are $3 \times 3 = 9$ of elements of this feature.

The final category of feature, *Ownership*, uses the fraction of time each team t has owned each domination location l during the time window δ to compute similarity. So, if x is the fraction of time t has controlled l in the current game and y is the analogous fraction from the case, then $Sim_{Own}(x,y) = 1 - |x-y|$. With two teams and three domination locations, this category has a total of six elements.

4 Knowledge Representation Requirements

The three agents have some common requirements: an API that allows them to sense information from the game world and sends commands to execute in the game world. All agents require a representation of the potential states of the world. The details of the representation vary from agent to agent but basically the state contains information about (1) which team owns each location, (2) where are the bots located, and (3) the score of the game. The agents also require information about the actions that the bots can make. This is basically a state-transition function $S \times A \rightarrow S$ that indicates for each state s and action a what next state will be reached if a is taken in s. Aside from these common requirements, some comparisons about each agent's knowledge representation requirements can be made (Table 3 summarizes the three systems).

Table 3 Knowledge requirements of systems

Game AI	Description
HTNBOTS	Action transition model represented as operators and HTN methods
RETALIATE	Action transition model
CBRETALITE	Action transition model, features for the cases, similarity metric

4.1 HTNBOTS *Has the Largest Knowledge Engineering Effort*

HTNBOTS uses the SHOP HTN planning algorithm [10]. SHOP uses a domain-independent algorithm to generate plans that are then executed as outlined in Sect. 3.1. In order to use SHOP, methods and operators must be provided. As explained in Sect. 3.1, methods encode how to achieve a compound task, whereas operators define valid actions in the domain, and methods provide knowledge about how to combine the actions to solve problems in the domain (Table 2 shows an example method and operator). A single operator can describe multiple transitions using variables; every possible instantiation of variables into constants is one possible transition. Hence, the collection of all operators encodes the state transition function, which as pointed out before is a common knowledge requirement for all agents. Creating methods requires a deep understanding of the domain to understand the ways in which problems can be solved. The difficulty of creating a list of methods to model completely a domain is a well-known limitation of HTN planning, and research has been conducted aiming at learning this knowledge automatically from a collection of sample plan traces [6, 17, 18].

4.2 RETALIATE *Has the Lowest Knowledge Engineering Effort*

RETALIATE only needs a state transition model (roughly equivalent to the operators in HTN planning) as input. For example, the operator from Table 2 is represented as multiple transitions of the form:

$$[s, \text{sendbot}(b, L_D), s'] \tag{2}$$

for every bot b and for every pair of states (s, s') such that s is a state where b is in L_C and s' is a state where b is in L_D. There is no knowledge about how to combine actions to solve problems because, as explained in Sect. 3.2, RETALIATE learns this knowledge as policies using Q-learning.

4.3 CBRETALITE *Has a Low Knowledge Engineering Effort*

In addition to the state transition model of RETALIATE, CBRETALITE needs to identify the features F that will be used to describe the problem section of cases.

There are many possible features of the game world that could be used, and selecting those that are most likely to contain useful information can be more of an art than a science. Additionally, an appropriate local similarity metric for each feature must be identified. Our global similarity metric is a weighted average of the local similarity metrics used for each feature, and setting these weights to an appropriate value is another knowledge engineering challenge.

To offset these burdens, we have also investigated a system to automate much of knowledge engineering process [5]. In that work, we include a large number of features and initially weight them all equally. Through an iterative process, we learn new weights that accurately represent the usefulness of the individual features and can remove those that have very low weights.

It is possible for a CBR system to be provided with a case library designed by experts, which would substantially increase the knowledge engineering effort required to use the system. While CBRETALITE will work with cases of any provenance, we have only used cases that it learns through its own experience. Thus, the case library is learned automatically and does not contribute to the knowledge engineering burden.

5 Game Performance Comparisons

To test the performance of our AI agents, we pitted them against a variety of hard-coded agents in the DOM game. The performance metric we used is the final score of the agent minus the final score of the opponent. This performance was measured in a variety of maps in a 3-bot versus 3-bot and 4-bot versus 4-bot settings. To determine the effectiveness of learning algorithms, we typically played several games against the same opponent sequentially, maintaining the knowledge base between them. The opponents were ranked among three classes of teams:

- *Easy-difficulty opponents*. The team encodes a simplistic strategy that is easy to counter.
- *Medium-difficulty opponents*. The team encodes a somewhat more difficult strategy to counter.
- *Hard-difficulty opponents*. The team's strategy is very difficult to counter.

This categorization was obtained through observation of multiple games across different games. The individual behavior of each bot was controlled by the same finite state machine; so this was not a factor for the difference in performance among the teams. We further define opponent strategy as either dynamic or static; dynamic opponents change their strategy over time while static opponents do not. Table 4 presents a summary of the results. For details please refer to the individual papers [1, 3, 7, 9, 16].

Some of the easy-difficulty opponents distributed bots among domination points in a fixed, hard-coded strategy. Others intelligently selected a subset of domination locations to contend for ordering one bot to defend each and any remaining bots to

Table 4 Performance of the three systems

Game AI	Description
HTNBOTS	Solid performance versus easy- and medium-difficulty opponents. It loses versus hard-difficulty opponents
RETALIATE	Solid performance versus easy- and medium-difficulty opponents. It wins versus some of the hard-difficulty opponents but loses to others
CBRETALITE	Improves the performance of RETALIATE against easy- and medium-difficulty opponents.

patrol between them. Medium-difficulty opponents used more complex strategies, such as always sending each bot to the unowned domination point that it is closest to, if any such points exist, or distributing the bots evenly among the domination points that it does not own, without using any to defend the points that it does own. The hard-difficulty opponents used even more complex strategies or dynamically selected among strategies based on the current situation.

5.1 HTNBOTS *Has a Solid Performance Versus the Easy and Medium Static Opponents*

HTNBOTS was the first system we built to play DOM games. Initially we had a relatively small set of opponents, which over time we found to be easy- and medium-difficult opponents. Against these the initial knowledge base did well. As we added more competent opponents over the years, the performance of HTNBOTS was poor even against some of the more recent medium-difficulty opponents. This was the result of lack of experience with the DOM game, which meant that the first encodings we created were a bit naïve. Against the most difficult ones, the performance of HTNBOTS is not as good as the other two agents. We also played HTNBOTS against RETALIATE and CBRETALITE directly, and it was usually defeated. This revealed some shortcomings in its knowledge base. Recently, the knowledge base of HTNBOTS went through a major overhaul, resulting in significant performance improvement [7, 9]. It now solidly beats all easy- and medium-difficulty opponents from our latest testbed. It is still outperformed by RETALIATE and CBRETALITE when competing versus the most difficult opponents. We believe that further improvements are attainable by modifying the existing HTN methods.

Over time, RETALIATE *achieves a solid performance versus the easy- and medium- static opponents, versus dynamic opponents, and versus some of the most difficult opponents.*

RETALIATE was the second game agent we created to play DOM games, and we benefitted from our experiences with HTNBOTS. In particular, we identified a small number of features, such as ownership of the domination locations, that are crucial representatives of the state of the game world. Against easy-difficulty opponents, RETALIATE quickly achieves a good performance (typically very early in the first game). Against medium-difficulty opponents, it will learn a winning policy within the first half of the game. Against some of the difficult opponents, it will still learn a winning policy within the first game. However, against others of the hard-difficulty opponents, it does not seem to be able to converge to a winning policy. We believe that the main factor for this latter behavior is that the current set of features selected is not sufficient to capture all necessary conditions that would allow RETALIATE to counter the strategies of these very difficult opponents. Nevertheless, it is remarkable that RETALIATE is able to learn a winning policy versus most opponents within one game. We also tested RETALIATE versus dynamic opponents (i.e., opponents that change their strategy over time), and it was able to adapt versus these opponents as well.

5.2 CBRETALITE *Improves the Performance Over* RETALIATE *on Easy- and Medium-Difficulty Opponents*

CBRETALITE was conceived with the idea of improving on the shortcomings of RETALIATE. Specifically, we expected it to find a winning policy faster than RETALIATE by short-circuiting the RL process in that it would immediately retrieve a "good" policy from the case library and, hence, void the need for RETALIATE to find a winning policy from scratch. For most of the easy- and medium-difficulty opponents, CBRETALITE was able to find a winning policy quicker than RETALIATE. This was less so versus the more difficult opponents. In some cases, it was able to improve on RETALIATE but the difference between the two was not statistically significant. We believe that the reason for these results is the same as the reason why RETALIATE cannot converge to a winning policy versus some of the hard opponents. Namely, that there are not enough features represented in the state to identify certain situations in the game world where taking one action over another one would be desirable. Still, we found the results promising as CBR helps to address one of the most significant shortcomings of RL by reducing the time the RL algorithm takes to converge to a winning policy.

6 Final Remarks

Game AI has received a lot of attention over recent years. There have been a number of works showcasing the use of AI techniques such as RL, planning, and CBR. With very few exceptions, including the Bridge Baron® and F.E.A.R.® systems, both of which use AI planning techniques [11, 13], there have been very few fielded applications of these techniques into modern commercial games. Our work showcases some of these difficulties:

- *Difficulty in creating the knowledge bases.* Developing competent players using deliberative reasoning such as HTN planning can require a significant effort to create the knowledge bases. This is consistent with observations about using HTN planning in other domains.
- *Time to generate competent policies.* Learning algorithms such as RL require some time until they converge to competent policies.

At the same time our study points to some promising capabilities. The crucial point is that game researchers have pointed out the need to create competitive AI rather than the most perfect possible one [12, 14]. The goal for common commercial game applications is, after all, not to create an AI that will become unbeatable for a human player but one that is competitive for an average player. With this goal in mind, our study points to the following possibilities:

- *Capability to create competent AI.* It is feasible to create good AI with deliberative reasoning techniques such as HTN planning. In fact our experiments show that even the first, somewhat naïve version of the knowledge base could beat all easy-difficulty and some of the medium-difficulty opponents. Further work was sufficient to create a competitive version that could only be beaten by the hard-difficulty opponents.
- *Capability to learn competent AI.* It is feasible to learn good AI within reasonable time. Albeit it requires a careful analysis of the features of the state of the game to identify a small subset of these features that is sufficient to guarantee good performance. In addition, CBR can further improve the speed upon which good performance is achieved by skipping several trial-and-error iterations.

For future work, the results of our study points toward an intriguing direction: deliberative AI such as HTN planning could be combined with learning techniques such as combining CBR and RL techniques to attain competent Game AI. Each could be used to address the shortcomings of the other one. HTN planning can start with somewhat competent game AI. This will reduce the knowledge engineering effort compared to creating a knowledge base that is fully competent. At the same time, it guarantees a minimum performance level from the outset of the game unlike RL. Using learning techniques one could improve the knowledge base to fill it with newly discovered strategies that the learning algorithm finds while playing. This would address the shortcoming of HTN planning where the initial knowledge base may encode some flawed strategies. There is a challenge with this direction, which is how to combine the symbolic plan generation process of HTN planning with the stochastic mechanism of RL. In recent work [4], we have begun initial work toward this combination.

Acknowledgements This work was sponsored by National Science Foundation (grant #0642882). The views, opinions, and findings contained in this chapter are those of the authors and should not be interpreted as representing the official views or policies, either expressed or implied, of the NSF.

References

1. Auslander, B., Lee-Urban, S., Hogg, C., Muñoz-Avila, H.: Recognizing the enemy: Combining reinforcement learning with strategy selection using case-based reasoning. In: Proceedings of the Ninth European Conference on Case-Based Reasoning (ECCBR-08), pp. 59–73. Springer (2008). Trier, Germany
2. Gillespie, K., Karneeb, J., Lee-Urban, S., Muñoz-Avila, H.: Imitating inscrutable enemies: Learning from stochastic policy observation, retrieval and reuse. In: Proceedings of the 18th International Conference on Case-Based Reasoning (ICCBR-10). AI Press (2010)
3. Hoang, H., Lee-Urban, S., Muñoz-Avila, H.: Hierarchical plan representations for encoding strategic game AI. In: Proceedings of the First Conference on Artificial Intelligence and Interactive Digital Entertainment (AIIDE-05). AAAI Press (2005). Marina del Ray, CA
4. Hogg, C., Kuter, U., Muñoz-Avila, H.: Learning methods to generate good plans: Integrating HTN learning and reinforcement learning. In: Proceedings of the Twenty-Fourth AAAI Conference on Artificial Intelligence (AAAI-10). AAAI Press (2010)
5. Hogg, C., Lee-Urban, S., Auslander, B., Muñoz-Avila, H.: Discovering feature weights for feature-based indexing of q-tables. In: Proceedings of the Uncertainty and Knowledge Discovery in CBR Workshop at the 9th European Conference on Case-Based Reasoning (ECCBR-08) (2008)
6. Hogg, C., Muñoz-Avila, H., Kuter, U.: HTN-Maker: Learning HTNs with minimal additional knowledge engineering required. In: Proceedings of the Twenty-Third AAAI Conference on Artificial Intelligence (AAAI-08). AAAI Press (2008)
7. Muñoz-Avila, H., Aha, D.W., Jaidee, U., Klenk, M., Molineaux, M.: Applying goal directed autonomy to a team shooter game. In: Proceedings of the Twenty-Third Florida Artificial Intelligence Research Society Conference (FLAIRS-10), pp. 465–470. AAAI Press (2010). Daytona Beach, FL
8. Muñoz-Avila, H., Hoang, H.: Coordinating teams of bots with hierarchical task network planning. In: S. Rabin (ed.) AI Game Programming Wisdom 3. Charles River Media, Boston, MA (2006)
9. Muñoz-Avila, H., Jaidee, U., Aha, D.W., Carter, E.: Goal directed autonomy with case-based reasoning. In: Proceedings of the Eighteenth International Conference on Case-Based Reasoning (ICCBR-2010). Springer (2010). Berlin
10. Nau, D.S., Cao, Y., Lotem, A., Muñoz-Avila, H.: SHOP: Simple hierarchical ordered planner. In: Proceedings of the Sixteenth International Joint Conference on Artificial Intelligence (IJCAI-99), pp. 968–973. AAAI Press (1999). Stockholm
11. Orkin, J.: Three states and a plan: The A.I. of F.E.A.R. In: Proceedings of the Game Developer's Conference (GDC) (2006)
12. Scott, B.: The illusion of intelligence. In: AI Game Programming Wisdom. Charles River Media (2002). Boston, MA
13. Smith, S.J.J., Nau, D., Throop, T.A.: A planning approach to declarer play in contract bridge. Computational Intelligence **12**(1), 106–130 (1996)
14. Spronck, P., Sprinkhuizen-Kuyper, I., Postma, E.: Difficulty scaling of game AI. In: A.E. Rhalibi, D.V. Welden (eds.) Proceedings of the 5th International Conference on Intelligent Games and Simulation (GAME-ON-04) (2004). EUROSIS, Belgium
15. Sutton, R.S., Barto, A.G.: Reinforcement Learning: An Introduction. MIT Press, Cambridge, MA (1998)
16. Vasta, M., Lee-Urban, S., Muñoz-Avila, H.: RETALIATE: Learning winning policies in first-person shooter games. In: Proceedings of the Seventeenth Innovative Applications of Artificial Intelligence Conference (IAAI-07). AAAI Press (2007)
17. Yang, Q., Wu, K., Jiang, Y.: Learning action models from plan traces using weighted MAX-SAT. Artificial Intelligence Journal (AIJ) **171**(2–3) (2007)
18. Zhuo, H.H., Hu, D.H., Hogg, C., Yang, Q., Muñoz-Avila, H.: Learning HTN method preconditions and action models from partial observations. In: Proceedings of the Twenty-first International Joint Conference on Artificial Intelligence (IJCAI-09). AAAI Press (2009)

Case-Based Reasoning and User-Generated Artificial Intelligence for Real-Time Strategy Games

Santiago Ontañón and Ashwin Ram

Abstract Creating artificial intelligence (AI) for complex computer games requires a great deal of technical knowledge as well as engineering effort on the part of game developers. This chapter focuses on techniques that enable end-users to create AI for games without requiring technical knowledge using case-based reasoning (CBR) techniques. AI creation for computer games typically involves two steps: (a) generating a first version of the AI, and (b) debugging and adapting it via experimentation. We will use the domain of real-time strategy games to illustrate how CBR can address both steps.

1 Introduction

Over the last 30 years computer games have become much more complex, offering incredibly realistic simulations of the real world. As the realism of the virtual worlds that these games simulate improves, players also expect the characters inhabiting these worlds to behave in a more realistic way. Thus, game developers are increasingly focusing on developing the intelligence of these characters. However, creating artificial intelligence (AI) for modern computer games is both a theoretical and engineering challenge. For this reason, it is hard for end-users to customize the AI of games in the same way they currently customize graphics, sound, maps, or avatars.

This chapter focuses on techniques to achieve *user-generated AI*, i.e., on techniques which would enable end-users to author AI for games. This is a complex task, since modern computer games are very complex. For example, real-time strategy (RTS) games (which will be the focus of this chapter) require complex strategic reasoning which includes resource handling, terrain analysis, or long-term planning under severe real-time constraints and without having complete information.

S. Ontañón (✉)
Artificial Intelligence Research Institute (IIIA-CSIC), Campus UAB, 08193 Bellaterra, Spain
e-mail: santi@iiia.csic.es

P.A. González-Calero and M.A. Gómez-Martín (eds.), *Artificial Intelligence for Computer Games*, DOI 10.1007/978-1-4419-8188-2_5,

Because of all of these reasons, programming AI for RTS games is a hard problem. Thus, we would like to allow end-users to create AI without programming.

When a user wants to create an AI, the most natural way to describe the desired behavior is by demonstration. Just let the user play a game demonstrating the desired behavior of the AI. Therefore, a promising solution to this problem is *learning from demonstration* (LfD) techniques. However, LfD techniques have their own limitations, and, given the complexity of RTS games and the lack of strong domain theories, it is not possible to generate an AI by generalization of a few human demonstrations.

The first key idea presented in this chapter is to use case-based reasoning (CBR) [1, 9] approaches for LfD. While it is hard to completely generalize an AI from a set of traces, it is possible to break demonstrations into smaller pieces, which contain specific instances of how the user wants the AI to behave in different situations. For instance, from a demonstration, the sequence of actions the user has used in a specific scenario to destroy an enemy tower can be extracted. These pieces correspond to what in CBR are called *cases*, i.e., concrete problem-solving episodes. Each case contains the actions the user wants the AI to perform in a concrete specific situation. Moreover, it is also possible to adapt cases to similar situations. Using a CBR approach to LfD, we do not need to completely generalize a demonstration. It is enough with being able to adapt pieces of it to similar situations. Moreover, as we will see, classic CBR frameworks need to be extended to deal with this problem. In order to illustrate these ideas, we will introduce a system called *Darmok 2* (D2), which is capable of learning how to play RTS games through LfD.

The second key idea presented in this chapter is that when creating AIs, either using LfD or directly coding them, it is very hard to achieve the desired result in the first attempt. Thus, using *self-adaptation* techniques, given a particular AI, it can be automatically adapted fixing some issues it might contain, or making it ready for an unforeseen situation. Again, self-adaptation is a hard problem because of two main reasons: first, how to detect that something needs to be fixed, and second, once an issue has been identified, how to fix it. We will see how this problem can again be addressed using CBR ideas, and specifically we will present a meta-reasoning approach inspired in CBR that addresses this problem. The main idea is to define a collection of *failure-patterns* (which could be seen as cases in a CBR system), that capture which failures to look for and how to fix them. In order to illustrate this idea, we will introduce the *Meta-Darmok* system, which uses meta-reasoning to improve its performance at playing RTS games.

In summary, the main idea of this chapter is the following. Authoring AI typically requires two processes: (a) creating an initial version of the AI and (b) debugging it. LfD is a natural way to help end-users with (a), and self-adaptation techniques can help users with (b). Moreover, both LfD and self-adaptation are challenging problems with many open questions. CBR can be used to address many of these open questions and thus make both learning from demonstration and self-adaptation feasible in the domain of complex computer games such as RTS games.

The remainder of this chapter is organized as follows. Section 2 very briefly introduces CBR. Sections 3 and 4 contain the main technical content of the chapter.

Section 3 focuses on CBR techniques for LfD in RTS games, and Sect. 4 focuses on CBR-inspired meta-reasoning techniques for self-adaptation. Section 5 concludes the chapter and outlines open problems to achieve user-generated AI.

2 Case-Based Reasoning

CBR [1, 9] is a problem-solving methodology based on reusing specific knowledge of previously experienced and concrete problem situations (*cases*). Given a new problem to solve, instead of trying to solve the problem from scratch, a CBR system will look for similar and relevant cases in its *case base*, and then adapt the solutions in these cases to the problem at hand. A typical case in a CBR system consists of a triple: problem, solution, and outcome, where the outcome represents the result of applying a particular solution to a particular problem.

The activity of a CBR system can be summarized in the CBR cycle, shown in Fig. 1, which consists of four stages: *Retrieve*, *Reuse*, *Revise*, and *Retain*. In the Retrieve stage, the system selects a subset of cases from the case base that are relevant to the current problem. The Reuse stage adapts the solution of the cases selected in the retrieve stage to the current problem. In the Revise stage, the obtained solution is examined by an oracle, which gives the correct solution (as in supervised learning). Finally, in the Retain stage, the system decides whether to incorporate the new solved case into the case base or not.

While inductive machine learning techniques learn from sets of examples by constructing a global model (a decision tree, a linear discrimination function, etc.) and then forgetting the examples, CBR systems do not attempt to generalize the

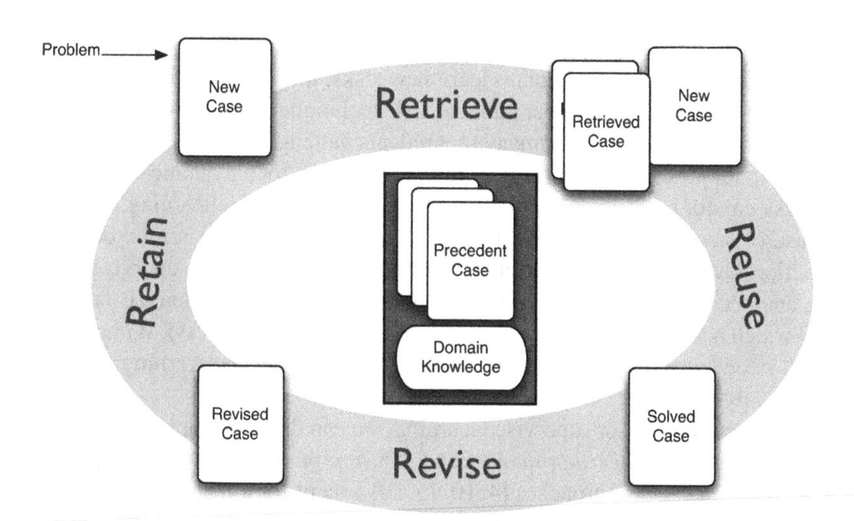

Fig. 1 The case-based reasoning cycle

cases they learn. CBR aligns with the ideas of *lazy learning* [2] in machine learning, where all kinds of generalization are performed at problem-solving time (during the Reuse stage). Thus, CBR systems only need to perform the minimum amount of generalization required to solve the problem at hand. As we will see, this is an important feature, since, for complex tasks like RTS games, attempting to learn a complete model of how to play the game by generalizing from a set of examples might be unfeasible.

3 Generating AI by Demonstration

A promising technology to achieve user-generated AI is *LfD* [20]. The goal of LfD is to learn how to perform a task by observing an expert. In this section, we will first introduce the main ideas of LfD, with a special emphasis on case-based approaches. Then we will explain how can they be applied to achieve user-generated AI by explaining how this is solved in the D2 system, which has been used to power a social gaming website, Make ME Play ME (MMPM), based around the idea of user-generated AI.

3.1 Background

LfD (also known as *programming by demonstration* or *programming by example*) has been widely studied in AI since early times [4] and specially in robotics [11] where lots of robotics-specific algorithms for learning movements from human demonstrations have been devised [14]. The main motivation behind LfD approaches is that learning a task from scratch, without any prior knowledge, is a very hard problem. When humans learn new tasks, they extract initial biases from instructors or by observing other humans. LfD techniques aim at imitating this process. However, LfD also poses many theoretical challenges.

LfD techniques typically attempt at learning a policy for a dynamic environment. This task cannot be addressed directly with inductive machine learning techniques because of several reasons: first, the performance metric might not be defined at the action level (i.e., we cannot create examples to learn using supervised learning); and second, we have the temporal blame assignment problem (it is hard to know which actions to blame or reward in case of failure or success). Without background knowledge, as evidenced by research in reinforcement learning, there is a prohibitively large space to explore.

In the same way as for supervised learning, we can divide the approaches to LfD in two large groups: *eager approaches* and *lazy approaches*, although work on LfD has focused on eager approaches [4, 10, 15, 20] except for a handful of exceptions like [8]. Eager methods aim at synthesizing a strategy, policy, or program, where as lazy approaches simply store the demonstrations (maybe with some preprocessing), and only attempt to generalize when facing a new problem. Let us present some representative work of LfD.

Tinker [10] is a programming by demonstration system, which could write arbitrary Lisp programs (containing even conditionals and recursion). The user provides examples as input/output pairs, where the output is a sequence of actions, and Tinker generalizes those examples to construct generic programs. Tinker allows the user to build incrementally, providing first simple examples and then move on to more complex examples. When Tinker needs to distinguish in between two situations, it prompts the user to provide a predicate that would distinguish them. Tinker is a classic example of an eager approach to LfD, where the system is trying to completely synthesize a program from the examples. Other eager approaches to LfD have been developed both in abstract AI domains [4] and in robotics domains [15].

In Tinker, we can already see one of the recurring elements in LfD systems: *traces*. A trace is the computer representation of a demonstration. It usually contains the sequence of actions that the user executed to solve a given problem. Thus, a pair problem/trace constitutes a demonstration, which is equivalent to a training example in supervised learning.

Schaal [20] studied the benefits of LfD in the context of reinforcement learning. He showed that under certain circumstances, the Q-value matrix can be primed using the data from the demonstration and achieved better results than a standard approach. This priming of the value matrix is a way to use the knowledge in the demonstrations to bias subsequent learning, and thus avoid blind search of the search space of policies. However, not all reinforcement learning approaches benefited from using the knowledge in the demonstrations. Note, moreover, that reinforcement learning also falls into the eager LfD approaches category, since it tries to obtain a complete policy.

Schaal's work evidences another of the important aspects in LfD: not all machine learning techniques easily benefit from the knowledge contained in the demonstrations.

In this chapter, however, we will focus on lazy approaches to LfD, based on CBR, which are characterized for not attempting to learn a general algorithm or strategy from demonstration, but at storing the demonstrations in some minimally generalized form to then adapt them to solve new problems. Other researchers have pursued similar ideas, like the work of Floyd et al. [8], which focuses on learning to imitate RoboCup players. Lazy approaches to LfD are interesting, since they can potentially avoid the expensive exploration of the large search space of programs or strategies. While the central problem of eager LfD approaches is how to *generalize* a demonstration to form a program, the central problem of lazy LfD approaches becomes how to *adapt* a demonstration to a new problem.

In order to apply LfD to a given task, several problems have to be addressed: how to generate demonstrations, how to represent each demonstration (trace) , how to segment demonstrations (which parts demonstrate which tasks and subtasks), which information to extract from the demonstrations, and how this information will be used by the learning algorithm. The remainder of this section will focus on a lazy LfD approach to learn AI in the context of computer games, and on how to address the issues mentioned above.

3.2 Learning from Demonstration in Darmok 2

D2 [16] is a real-time case-based planning [21] system designed to play RTS games. D2 implements the *on-line case-based planning* cycle (OLCBP) as introduced in [17]. The OLCBP cycle attempts to provide a high-level framework to develop case-based planning systems that operate online, i.e., that interleave planning and execution in real-time domains. The OLCBP cycle extends the traditional CBR cycle by adding two additional processes, namely *plan expansion* and *plan execution*. The main focus of D2 is to explore learning from unannotated human demonstrations and the use of adversarial planning techniques. In this section we will focus on the former.

3.2.1 Representing Demonstrations, Plans, and Cases

A demonstration in D2 is represented as a list of triples $[\langle t_1, G_1, A_1 \rangle, \ldots, \langle t_1, G_n, A_n \rangle]$, where each triple contains a time stamp t_i game state G_i and a set of actions A_i (that can be empty). The set of triples represent the evolution of the game and the actions executed by each of the players at different time intervals. The set of actions A_i represent actions that were issued at t_i by any of the players in the game. The game state is stored using an object-oriented representation that captures all the information in the state: map, players, and other entities (entities include all the units a player controls in an RTS game: e.g., tanks).

Unlike in traditional STRIPS planning [7], actions in RTS games may not always succeed, they may have nondeterministic effects, and they might not have an immediate effect, but be durative. Moreover, in a system like D2, it is necessary to be able to monitor executing actions for progress and check whether they are succeeding or failing. Thus, a typical representation of preconditions and postconditions is not enough. An action a is defined in D2 as a tuple containing seven elements including success conditions and failure conditions [16]. However, for the purposes of LfD, precondition and postcondition suffice.

Plans in D2 are represented as *hierarchical petri nets*. Petri nets [13] offer an expressive formalism for representing plans that include conditionals, loops, or parallel sequences of actions. In short, a petri net is a graph consisting of two types of nodes: *transitions* and *states*. Transitions contain conditions and link states to each other. Each state might contain *tokens*, which are required to fire transitions. The flow of tokens in a petri net represents its status. In D2, the plans that will be learned by observing demonstrations consist of hierarchical petri nets, where some states will be associated with sub-plans, which can be primitive actions or

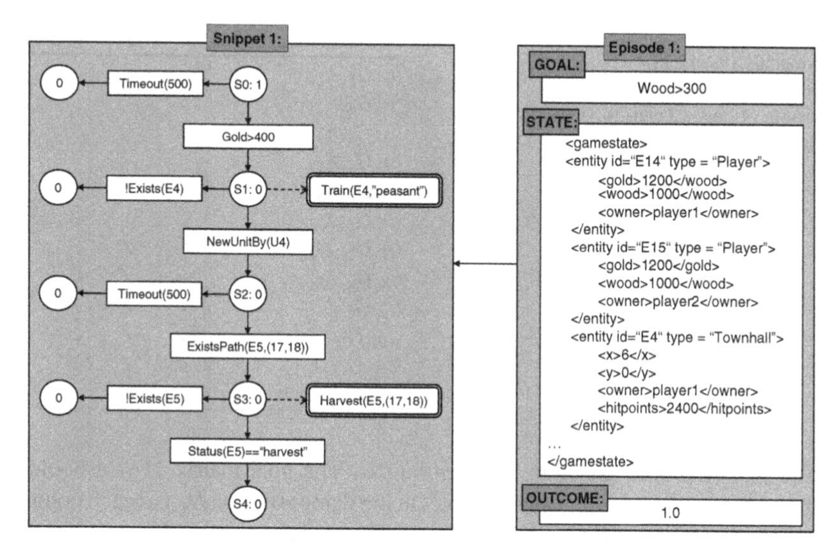

Fig. 2 A case in D2 consisting of a snippet and an episode. The snippet contains two actions, and the episode says that this snippet succeeded in achieving the goal $Wood > 300$ in the specified game state. The game state representation is not fully included due to space limitations

subgoals. The left-hand side of Fig. 2 shows an example of a petri net representing a plan consisting of two actions to be executed in sequence: *Train(E4, "peasant")* and *Harvest[E5,(17,18)]*. Note that the handling of preconditions, postconditions, etc. is handled by the petri net, making the execution module of D2 a simple petri net simulation component.

When D2 learns plans from demonstrations, each plan is stored as a *case*. Cases in D2 are represented like cases in the Darmok system [17], consisting of a collection of plan *snippets* with *episodes* associated with them. As shown in Fig. 2, a snippet is a petri net, and an episode is a structure storing the outcome obtained when a particular snippet was executed in a particular game state intending to achieve a particular goal. The outcome is a real number in the interval $[0, 1]$ representing how well the goal was achieved: 0 represents total failure and 1 total success.

3.2.2 Learning Plans and Cases from Demonstration

D2's case base is populated by learning both snippets and episodes from human demonstrations. The input to the learning algorithm is one demonstration D (of length n), a player p (D2 will learn only from the actions of player p in the demonstration D), and a set of goals G for which to look for plans. The output is a collection of snippets and episodes. The set of goals G can be fixed beforehand for every particular domain and is equivalent to the list of *tasks* in the HTN planning framework (thus, the inputs are the same as for the HTN-Maker algorithm). The learning process of D2 can be divided into three stages: *goal matrix generation*, *dependency graph generation*, and *hierarchical composition*.

Table 1 Goal matrix for a set of five goals $\{g_1, g_2, g_3, g_4, g_5\}$ and for a small trace consisting of only 12 entries (corresponding to the actions shown in Fig. 3, $A_{12} = \emptyset$)

Demonstration	g_1	g_2	g_3	g_4	g_5
$\langle t_1, G_1, A_1\rangle$					
$\langle t_2, G_2, A_2\rangle$					
$\langle t_3, G_3, A_3\rangle$					
$\langle t_4, G_4, A_4\rangle$					
$\langle t_5, G_5, A_5\rangle$					
$\langle t_6, G_6, A_6\rangle$				✓	
$\langle t_7, G_7, A_7\rangle$			✓	✓	
$\langle t_8, G_8, A_8\rangle$		✓	✓	✓	
$\langle t_9, G_9, A_9\rangle$		✓	✓	✓	✓
$\langle t_{10}, G_{10}, A_{10}\rangle$		✓	✓	✓	✓
$\langle t_{11}, G_{11}, A_{11}\rangle$		✓	✓	✓	✓
$\langle t_{12}, G_{12}, A_{12}\rangle$	✓	✓	✓	✓	✓

The first step is to generate the *goal matrix*. The goal matrix M is a boolean matrix, where each row represents a triplet in the demonstration D, and each column represents one of the goals in G. $M_{i,j}$ is true if the goal g_j is satisfied at time t_i in the demonstration. An example goal matrix can be seen in Table 1.

Once the goal matrix is constructed, a set of *raw* plans P are extracted from it in the following way:

1. For each goal $g_j \in G$ do
   a. For each $0 < i \leq n$ such that $M_{i,j} \wedge \neg M_{i-1,j}$ do
      i. Find the largest $0 < l < i$ such that $\neg M_{l,j} \wedge (l = 1 \vee M_{l-1,j})$
      ii. Generate a raw plan from the actions executed by player p in the set $A_l \cup A_{l+1} \cup \ldots \cup A_{i-1}$ and add it to P

For example, five plans could be generated from the goal matrix in Table 1. One for g_1 with actions $A_l \cup \ldots \cup A_{12}$, one for g_2 with actions $A_l \cup \ldots \cup A_8$, one for g_3 with actions $A_l \cup \ldots \cup A_7$, one for g_4 with actions $A_l \cup \ldots \cup A_6$, and one for g_5 with actions $A_l \cup \ldots \cup A_9$. Note that the intuition behind this process is just to look at sequences of actions that happened before a particular goal was satisfied, since those actions are a plan to reach that goal. Many more plans could be generated by selecting subsets of those plans, but since D2 works under tight real-time constraints, currently it learns only a small subset of plans from each demonstration.

Note that this process is enough to learn a set of raw plans for the goals in G. The snippets will be constructed from the aforementioned sets of actions, and the episode will be generated by taking the game state in which the earliest action in a particular plan was executed. Note that all plans extracted using this method are plans that succeeded; thus all episodes have outcome equal to 1. However, these raw plans might contain unnecessary actions and would be monolithic, i.e., they will not be decomposable hierarchically into subgoals. Dependency graph generation and hierarchical composition are used to solve both problems.

Given a plan consisting of a partially ordered collection of actions, a *dependency graph* [24] is a directed graph where each node represents one action in the plan, and edges represent dependencies among actions. Such a graph is used by D2 to remove unnecessary actions from the learned plans.

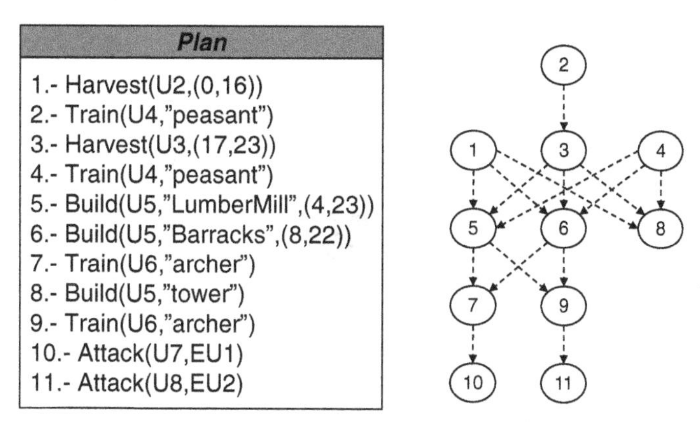

Fig. 3 An example dependency graph constructed from a plan consisting of 11 actions in an RTS game

Such a graph is easily constructed by checking each pair of actions a_i and a_j in the plan, and checking first of all whether there is any order restriction between a_i and a_j. Only those pairs for which a_i can happen before a_j will be considered. Next, if one of the postconditions of a_i matches any precondition of a_j, and there is no action a_k that has to happen after a_i that also matches with that precondition, then an edge is drawn from a_i to a_j in the dependency graph, annotating it with which is the pair of postcondition/precondition that matched. Figure 3 shows an example dependency graph (where the labels in the edges have been omitted for clarity). The plan shown in the figure shows how each action is dependent on each other, and it is useful to determine which actions contribute to the achievement of particular goals.

D2 constructs a dependency graph of the plan resulting from using the complete set of actions that a player p executed in a demonstration D. This dependency graph will be used to remove unnecessary actions from the smaller raw plans learned from the goal matrix in the following way:

1. For each plan $p \in P$ do
   a. Extract the subgraph of the dependency graph containing only the actions in p
   b. Detect which is the subset of actions A from the actions in p such that their postconditions match with the goal of plan p
   c. Remove from p all actions that according to the subgraph do not contribute directly or indirectly to any of the actions in A

Moreover, the plan graph provides additional internal structure to the plan, indicating which actions can be executed in parallel, and which ones have to be executed in a sequence. All this information is exploited when generating the petri net corresponding to the plan.

Finally, D2 analyzes the set of plans P resulting from the previous step using the dependency graph to see whether any of those plans are a sub-plan of another plan. Given two plans $p_i, p_j \in P$, if the set of actions in p_i is a subset of the set of

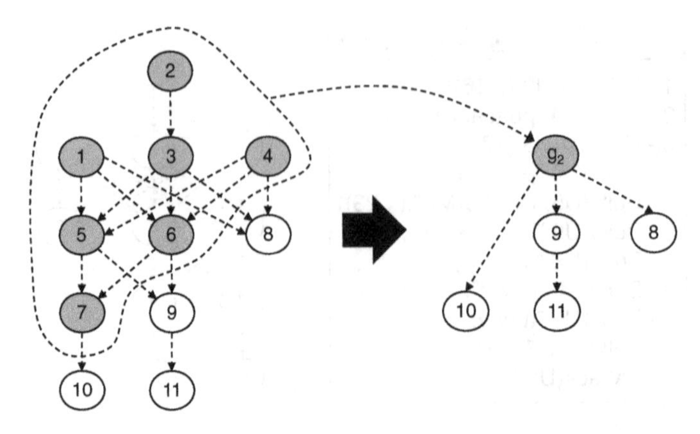

Fig. 4 The nodes greyed out in the *left* dependency graph correspond to the actions in the plan learned from a goal g_2, after substituting those actions by a single subgoal, the resulting plan graph looks like the one on the *right*

actions in p_j, D2 assumes that p_i is a sub-plan of p_j, and all the actions in p_i also contained in p_j are substituted by a single subgoal in p_j. Converting flat plans into hierarchical ones is important in D2, since it allows D2 to combine plans learned from one demonstration with plans learned from another at run time, increasing its flexibility.

Figure 4 shows an example of this process taking the plan graph of the plan learned for goal g_1 in Table 1, and substituting some of its actions by a single subgoal g_2. The actions marked in gray in the left-hand side of Fig. 4 correspond to the actions in the plan learned for g_2.

Note that the order in which we attempt to substitute actions by subgoals in plans will result in different final plans. Currently, D2 uses the heuristic of attempting first to substitute larger plans first. However, this issue is a subject of our ongoing research effort. Let us explain how can D2 be used for achieving user-generated AI.

Finally, it is worth to remark that D2's goal is not to learn how to play the game in an optimal way, but to learn the player's strategy. In this sense, it differs from other LfD strategies. For example, the techniques presented by Schaal [20] used LfD only to bias the learning process, which would proceed then to optimize the strategy using standard reinforcement learning.

3.3 Using Darmok 2 for User-Generated AI: Make ME Play ME

Make ME Play ME[1] is a project to build a social gaming website (see Fig. 5) based on the idea of user-generated AI and powered by D2. In MMPM, users do not just

[1] `http://www.makemeplayme.com`

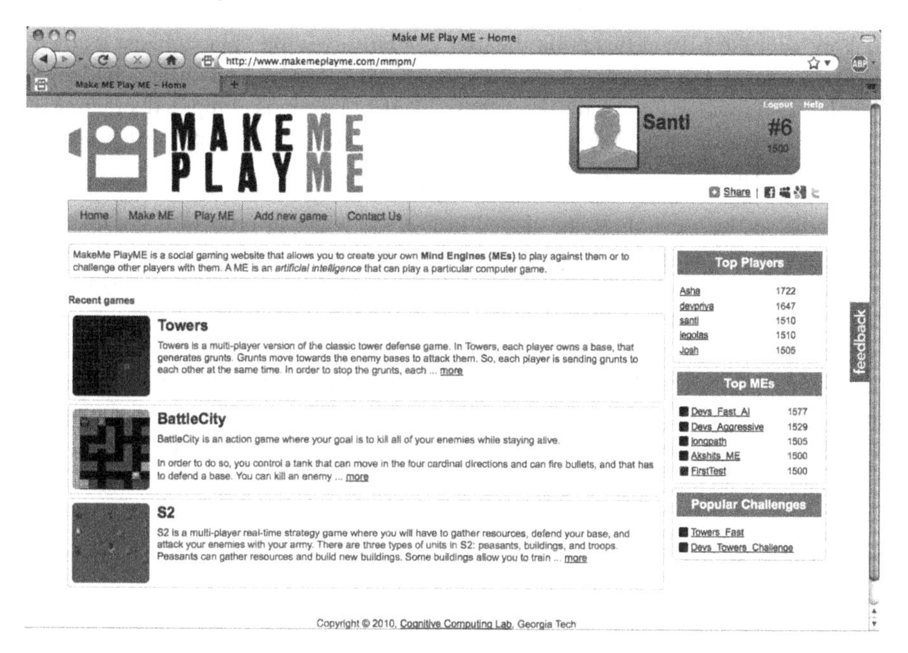

Fig. 5 The game selection page of Make ME Play ME

play games, they create their own AIs, called *Mind Engines* (MEs). Users train their own MEs, which can play the different games available in the website, and compete against the MEs created by other players. MMPM is not the first web or game where users can create their own AIs and make them compete with others, but it is the first one where users can create their own AIs by demonstration: users do not require programming knowledge, they just have to play a series of games demonstrating the strategy they want their ME to use.

In order to make user-generated AI a reality, many user interaction problems need to be addressed in addition to the technical problems concerning LfD explained in the previous section: for instance, how to generate demonstrations, or how to visualize the result of learning. In our work on MMPM, we focused on the first of these problems. The latter is still subject of our future work.

The user flow works as follows:

1. Play demonstration games: The user selects a game, configures it (selecting number of players, opponents, map, etc.), and then simply plays. The user can repeat this process as many times as desired. For each game played, a trace will be automatically saved by MMPM.
2. Create a ME: To create a ME, the user first selects which games does he wants to create a ME. Then MMPM lists the set of all available traces for that game (generated in the previous step). The user simply selects a subset of them (which will constitute the set of demonstrations), and the ME is created automatically, without further user intervention.

3. Play with the ME: at this point the user can already wither play against its own ME, or make the ME play with other users' MEs. MMPM lets users challenge other users' MEs. For each ME, a chess-like ELO score is computed, creating a leader-board of MEs. The users are thus motivated to create better MEs, which can climb up the leader boards.

Thanks to the technology developed in D2, the learning process is completely transparent to the user, who only needs to play games. There are no parameters that need to be set by the user. In order to achieve that, all the game-specific parameters of D2 are set before hand. When a new game is added to MMPM, the game creator is responsible for defining the goal ontology, and for specifying any other parameter that D2 needs to know about the game (e.g., whether the game is turn-based or real-time). Currently, MMPM hosts three different games, but more are on preparation, and it even has the functionality to allow users to upload their own games.

3.4 Discussion

MMPM and D2 allow users to author AIs simply by demonstrations. For instance, in previous work, we showed how it is easy to author an AI for the game Wargus (a clone of WARCRAFT II) by demonstration which can defeat the built-in AI [17]. Moreover, the resulting AIs clearly use the strategies demonstrated by the users. The learning process of D2 is efficient and learning does not take any perceptible time. Moreover, the planning algorithms of D2 are also efficient enough to work on real time in the set of games available in MMPM.

However, MMPM and D2 still display a number of limitations, some of which clearly correspond to open problems in LfD.

- First of all, the case-based planning approach of D2 is suitable for some kind of games (like RTS games), but breaks when the game becomes more reactive than deliberative. For example, one of the games in MMPM (BattleCity) is a purely reactive game, for which learning plans does not make much sense and where a more reactive approach like that in [8] should work much better.
- In addition to demonstrations, some LfD approaches also allow the user to provide feedback when the system performs the learned strategies to continue learning. In the context of D2 and computer games, it would be very valuable to allow such feedback, since it will enable the user to fine-tune the demonstrated strategies. However, this raises both technical and user-interface problems. The main technical problem is related to the delayed blame assignment problem: if the user provides a negative feedback, which of the previous decisions is to blame? Additionally, there would be user-interface problems that need to be solved about how can the user provide feedback on the actions being executed by the AI, specially in RTS games where a large number of actions is executed per second.

- Another issue, subject for our future research and common to all lazy learning approaches, is how to visualize the result of learning. Eager LfD techniques learn a policy or a program which can be displayed to the user in some form. But lazy LfD techniques do not. The only thing that could be displayed are the set of plans being learned. But that can be a potentially very large number of plans, and which does not include the procedure for selecting which plan to select in each situation (which is performed at run-time).
- Clearly, the biggest problem in LfD is how to generalize from a demonstration to a general strategy. Specifically, D2 is based on case-based planning, and this problem is translated into how can plans be adapted to new situations. This is a well-known problem in the case-based planning community [21] and has been widely studied. In D2 we used an approach with a collection of simplification assumptions which allow D2 to be able to adapt plans in real time [24]. However, those assumptions have been designed with RTS games in mind. Finding general ways to adapt plans in an efficient way for other game genres is still an open research issue.

4 Self-Adaptive AI Through Meta-Reasoning

Last section focused on techniques to easily generate AI for games. In this section, we are going to turn our attention to the complementary problem of how can AI self-adapt to fix any flaws that might have occurred during the learning process, or to adapt the AI to novel situations. This is known as the *adaptive-AI* problem in game AI. This section will provide a brief overview of the problem, and then focus on a solution which combines CBR with meta-reasoning, specifically designed for the problem of achieving user-generated AI in games.

4.1 Background

The most widely used techniques for authoring AI in commercial games are scripts [5] and finite-state machines [19] (and recently, behavior trees [18]). These techniques share one feature: once they have been authored, the behavior of the AI will be static: i.e., it will always be the same game after game (ignoring the trivial differences which can be introduced adding randomness). Static behavior can lead to suboptimal user experience, since, for instance, users might find a hole in the AI and exploit it continuously, or there might be an unpredicted situation or player strategy to which the AI does not know how to react. Trying to address this issue is known as achieving *adaptive game AI* [23].

Basically, adaptive game AI aims at developing techniques which allow for automatic self-modification of the game AI. A potential benefit is for fixing potential failures of the AI, but other uses have been explored, like using self-adaptation for automatically adjusting the difficulty level of games [22]. In this section, we are interested in the former and, specifically, in developing techniques which ease user-generated AI. Algorithms that enable self-adaptive AI would enable the users to create AI in an easier way, since some errors in their AI could be automatically fixed by the adaptive AI. Before presenting how CBR can be used to address this issue, let us briefly introduce some brief background and existing work.

Spronck et al. [23] identified a collection of requirements for adaptive game AI. Four are computational requirements: speed, effectiveness, robustness, and efficiency; and four are functional requirements: clarity, variety, consistency, and scalability. Some of those eight properties, however, apply only to *on-line* techniques for self-adaptation. Our interest in self-adaptive AI concerns allowing user-generated AI, and thus *off-line* adaptive AI techniques are also interesting. The most basic elements required to define adaptive AI are:

- *Representation of the AI*: a script, a collection of rules, a set of cases, etc.
- *Performance criteria*: if the AI has to be adapted, it is to improve in some measure. For instance, we might want to make the AI better, or better exhibit a particular strategy, or better adjust to the skill level of the player.
- *Allowed modifications*: which adaptations are allowed? Some times, adaptation simply means selecting among a set of given rule sets, some times, the rules or scripts can be actually adapted. This defines the space of possible adaptations.
- *Adaptation strategy*: which machine learning technique to use.

The most common approach to adaptive AI is letting the user define a collection of scripts or rules that define the behavior of the AI, and then learn which of those scripts or which subset of rules works better for each particular game situation according to a predefined performance criteria. This approach has been attempted both using reinforcement learning [23] and CBR [3].

Let us now present a technique that can be combined to the LfD techniques presented in the previous section, to ease the job of a user who wants to create an AI.

4.2 Automatic Plan Adaptation in Meta-Darmok

Meta-Darmok [12] is a case-based planning system based on the Darmok system [17], which is a predecessor to the D2 system described in the previous section. Meta-Darmok learns plans from expert demonstrations and then uses them to play games using case-based planning. Meta-Darmok is designed to play Wargus, and specially to automatically adapt Darmok's learned plans over time. The performance of Darmok, as well as D2, highly depends on the quality of the demonstrations provided by the user. If the demonstrations are poor, Darmok's behavior will be poor. If there is a mistake in one of the plans learnt from an expert trace, Darmok will repeat that mistake again and again each time Darmok retrieves that plan. The meta-reasoning approach presented in this section provides Darmok exactly with that capability, resulting in a system called Meta-Darmok, shown in Fig. 6.

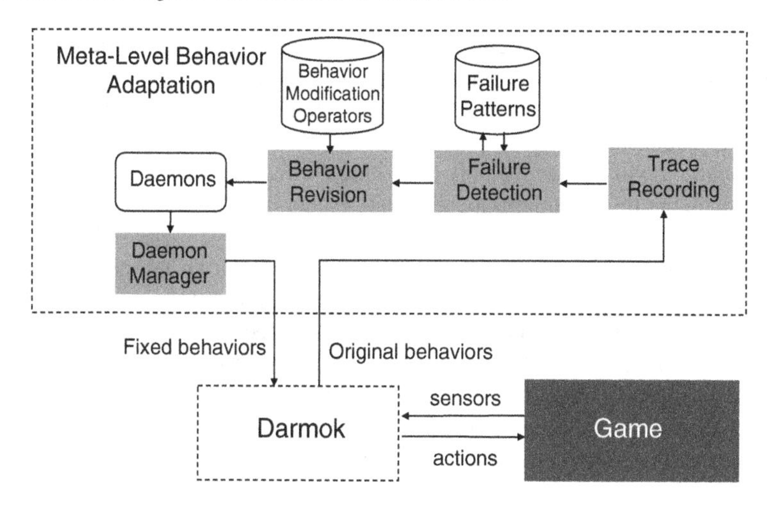

Fig. 6 Meta-reasoning flow of Meta-Darmok

Meta-Darmok's adaptation approach is based on the following idea: instead of fixing the plans one by one, a user can fix a collection of plans by defining a set of typical failures and associating a fix with them. Meta-Darmok's meta-reasoning layer constantly monitors the plans being executed to see whether any of the user-defined failures is happening. If failures occur, Meta-Darmok will execute the appropriate fixes. Moreover, Meta-Darmok's plan fixing happens off-line, after a game has been played. Note that this approach is radically different from approaches like reinforcement-learning, where the behavior is optimized by trial and error.

Specifically Meta-Darmok's approach consists of four parts: *Trace Recording*, *Failure Detection*, *Plan Modification*, and the *Daemon Manager*. During trace recording, a trace holding important events happening during the game is recorded. Failure detection involves analyzing the execution trace to find issues with the executing plans using a set of *failure patterns* [26]. These failure patterns capture the set of user-defined prototypical failures. Once a set of failures has been identified, the failed conditions can be resolved by appropriately revising the plans using a set of *plan modification routines*. These plan modification routines are created using a combination of basic modification operators (called *modops*, as explained later). Specifically, in Meta-Darmok, the modifications are inserted as *daemons*, which monitor for failure conditions to happen during execution when Darmok retrieves some particular plans; but in general, they could be implemented in a different way. A daemon manager triggers the execution of such daemons when required.

4.2.1 Trace Recording

While Meta-Darmok is playing a game, the trace recording module records an *execution trace*, which contains information related to basic events including the name of the plan that was being executed, the corresponding game state when the event occurred, the time at which the plan started, failed, or succeeded, and the delay from the moment the plan became ready for execution to the time when it actually started executing. The execution trace provides a considerable advantage in performing plan adaptation with respect to only analyzing the instant in which the failure occurred, since the trace can help localize past events that could possibly have been responsible for the failure.

Once a game finishes, an *abstracted trace* is created from the execution trace that Darmok generates. While the execution trace contains all the information concerning plan execution during the game, the abstracted trace contains only some key pieces of information: those needed to determine whether any failure pattern occurred during the game. The information included in the abstracted trace depends on which conditions are used in the failure patterns. For instance, for the set of patterns used in Meta-Darmok, information about hit points, location, actions being executed by the units, and in which cycles were units created or killed is included.

4.2.2 Failure Detection

Failure detection involves localizing the failures in the trace. Traces can be extremely large, especially in the case of complex RTS games on which the system may spend a lot of effort attempting to achieve a particular goal. In order to avoid the potentially very expensive search process of finding which actions are responsible for failures, the set of user-provided failure patterns can be used [6]. Failure patterns can be seen as a case-based approach to failure detection, and they greatly simplify the blame-assignment process into a search for instances of the particular problematic patterns.

Failure patterns are defined as finite state machines (FSMs) that look for generic patterns in the abstracted trace. An example of a failure pattern represented as FSM is *Very Close Resource Gathering Location failure* (VCRGLfail) (shown in Fig. 7)

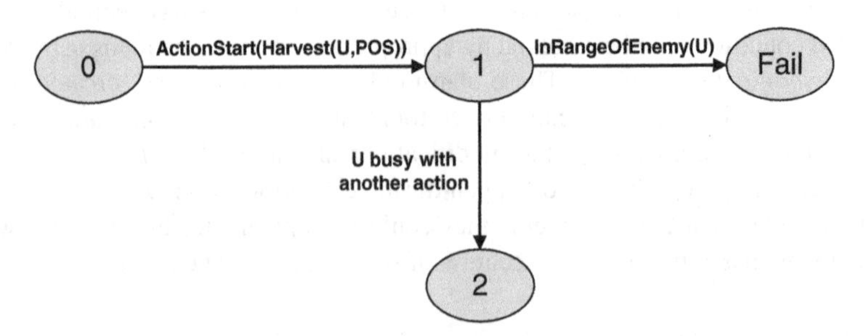

Fig. 7 FSM corresponding to the failure pattern VCRGLfail. This pattern detects a failure if the FSM ends in the *Fail* state. When a unit is ordered to start harvesting, the FSM moves to state 1, if the unit stops harvesting, it will move to state 2, and only when the unit gets in range of an enemy unit while harvesting, the FSM will end in the *Fail* state

Table 2 Some example failure patterns and their associated plan modification operators

Failure pattern	Plan modification operator
Resource idle failure (e.g., resource like peasant, building, enemy units could be idle)	Use the resource in a more productive manner (e.g., send peasant to gather more resources or use the peasant to create a building that could be needed later on)
Very close resource gathering location failure	Change the location for resource gathering to a more appropriate one
Inappropriate enemy attacked failure	Direct the attack toward the more dangerous enemy unit
Inappropriate attack location failure	Change the attack location to a more appropriate one
Basic operator failure	Adding a basic action that fixes the failed condition

that detects whether a peasant is gathering resources at a location that is too close to the enemy. This could lead to an opening for enemy units to attack early. Other examples of failure patterns and their corresponding plan modification operators are given in Table 2. Each failure pattern is associated with modification routines. When a failure pattern generates a match in the abstracted trace, an instantiation of the failure pattern is created. Each instantiation contains which were the particular events in the abstracted trace that matched with the pattern. This is used to instantiate particular plan modification routines that are targeted to the particular plans that were to blame for the failure.

4.2.3 Plan Modification

Once the cause of the failure is identified, it needs to be addressed through the appropriate modifications (modops). Modops can take the form of inserting or removing steps at the correct position in the failed plan, or changing some parameter of an executing plan. Each failure pattern has a sequence of modops associated with it. This sequence is called a *plan modification routine.*

Once the failure patterns are detected from the execution trace, the corresponding plan modification routines and the failed conditions are inserted as daemons for the plan in which these failed conditions are detected. The daemons act as a meta-level reactive plan that operates over the executing plan at runtime. The conditions for the failure pattern become the preconditions of the daemon, and the plan modification routine consisting of basic modops becomes the steps to execute when the daemon executes. The daemons operate over the executing plan, monitor their execution, detect whether a failure is about to happen, and repair the plan according to the defined plan modification routines.

Note that Meta-Darmok does not directly modify the plans in the case base of Darmok, but reactively modifies those plans when Darmok is executing them. In the current system, we have defined 20 failure patterns and plan modification routines for Wargus. The way Meta-Darmok improves over time is by accumulating the daemons that the meta-reasoner generates (which are associated with particular maps). Thus, over time, Meta-Darmok improves performance by learning which combination of daemons improves the performance of Darmok for each map. Using this approach, we managed to multiply by 2 the win-ratio of Darmok against the built-in AI of Wargus [12].

The adaptation system can be easily extended by writing other patterns of failure (as described in [25]) that could be detected from the abstracted trace and the appropriate plan modifications to the corresponding plans that need to be carried out to correct the failed situation.

4.3 Using Meta-Darmok for User-Generated AI

In order to use Meta-Darmok for user-generated AI, we integrated Meta-Darmok into a behavior authoring environment, which we call an *intelligent* IDE (iIDE). Specifically, we integrated authoring by demonstration, visualization of the behavior execution, and self-adaptation through meta-reasoning. The iIDE allows the game developer to specify initial versions of the required AI by demonstrating them instead of having to explicitly code the AI. The iIDE observes these demonstrations and automatically learns plans (that we will call *behaviors* in this section) from them. Then, at runtime, the system monitors the performance of these learned behaviors that are executed. The system allows the author to define new failure patterns on the executed behavior set, checks for predefined failure patterns, and suggests appropriate revisions to correct failed behaviors. This approach to allow definition of possible failures with the behaviors, detecting them at run-time, and proposing and allowing a fix selection for the failed conditions enables the author to define potential failures within the learnt behaviors and revise them in response to things that went wrong during execution.

Here we will focus only on how meta-reasoning is integrated into the iIDE (for more details about the iIDE reported here, see [25]). In order to integrate Meta-Darmok into the iIDE, we added to functionalities:

- *Behavior Execution Visualization and Debugging*: The iIDE presents the results of the executing behaviors in a graphical format, where the author can view their progress and change them. The author can also pause and fast-forward the game to whichever point he chooses while running the behaviors, make a change in the behaviors if required, and start it up again with the new behaviors to see the performance of the revised behaviors. The capability of the iIDE to fast-forward and start from a particular point further allows the author to easily replicate a possible bug late in the game execution and debug it. Figure 8 shows a screenshot of the execution visualization view in the iIDE, showing an executing behavior (including the current state of all the subgoals and actions).
- *Failure Detection and Fixing*: The iIDE authoring tool allows the author to visualize relevant chronological events from a game execution trace. The data allow

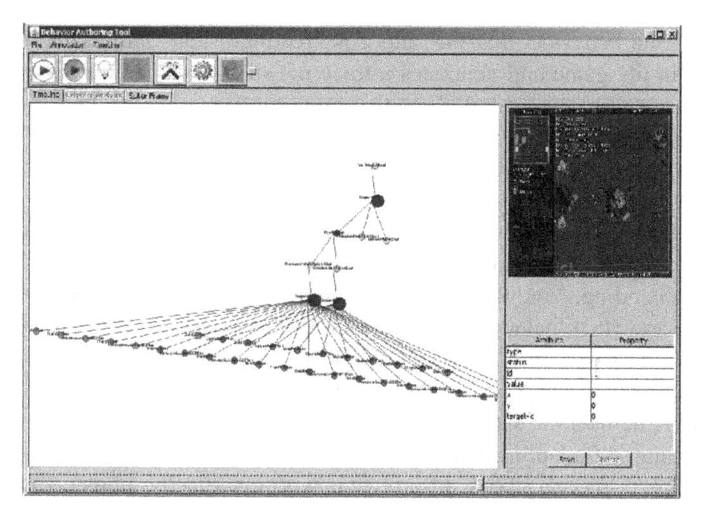

Fig. 8 A screenshot of the iIDE, showing the behavior execution visualization interface

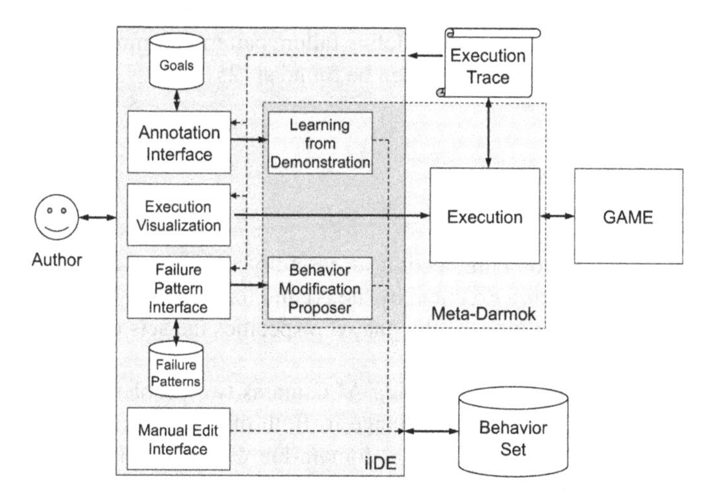

Fig. 9 Overview of how the iIDE interacts with the author and the game

the author define new *failure patterns* by defining combinations of these basic events and pre-existing failure conditions. Each failure pattern is associated with a possible fix. A fix is basically a proposed modification for a behavior that fixes the error detected by the failure pattern. When a failure pattern is detected, the iIDE suggests a list of possible fixes, from which the author can select an appropriate one to correct the failed behavior. These techniques were also previously developed by us in the context of believable characters [26].

Figure 9 shows an overview of how all the components fit together to allow the author to edit a proper behavior set for the game. The iIDE controls Meta-Darmok by sending the behaviors that the author is creating. Meta-Darmok then runs the behaviors in the game and generates a trace of what happened during execution. This trace is sent back to the iIDE so that proper information can be shown to the author. Basically, the iIDE makes the functionality of Meta-Darmok (LfD and self-adaptation through meta-reasoning) accessible to the user to allow easy behavior authoring.

We evaluated this iIDE with a small set of users, and the conclusions found that users felt authoring by demonstration was more convenient than writing our behaviors through coding. Note that it took not more than 25 min to generate behaviors to play Wargus (that includes the time to generate the demonstration playing plus trace annotation). However, since Meta-Darmok is based on the old Darmok system, which required trace annotation, users felt that annotation was difficult, since it was difficult to remember the actions they had performed.

Concerning self-adapting behaviors using meta-reasoning, users felt it was a very useful feature. However, they had problems with our specific implementation because the requirement that a failure pattern should occur inside the game to be able to define it was a setback. Users could think of simple failure patterns which they would like to add without having to even run the game. However, despite these problems, users were able to successfully define failure patterns. A more comprehensive explanation of the evaluation results can be found at [25].

4.4 Discussion

The techniques presented in this section successfully allow a system to detect problems in the behaviors being executed by the AI and fix them. However, we do so at the expense of letting the user be the one who specifies the sets of failures to look for, in the form of failure patterns.

Clearly, the problem of self-adapting AI contains two problems: detecting that something has to be changed, and change it. Both of them are, as of today, open problems. In our approach, we used a domain-knowledge intensive approach for detecting that something has to be changed, by letting the user specify domain-dependent failure patterns, which for the purposes of user-generated AI worked adequately, but at the expense of making the user having to manipulate concepts like conditions, actions, etc. when defining the failure patterns.

However, detecting that something has to be changed is a challenging problem. For example, in techniques such as dynamic scripting [23], we need to define a performance metric. In case the goal is just to adapt an AI to make it stronger or weaker, a performance metric is easy to define (percentage of wins, or some related measure should suffice). However, when the goal is to adapt an AI to better behave the way the user wants, this is harder, and interfaces to allow the user to provide feedback are required.

In general, for behavior creation, as we explained above, LfD is an intuitive way in which a user can provide domain knowledge. The iIDE presented in this section is an attempt to achieve the same thing for the problem of adapting an already created behavior. Other strategies that can be used are direct positive or negative reinforcement from the user when behaviors are being executed. This requires the user to constantly provide feedback, whereas failure patterns can be given once and be reused multiple times.

5 Conclusions

This chapter has focused on CBR techniques to achieve user-generated AI. We have presented two complementary techniques, LfD and self-adaptation, which when combined can help the task of an end-user who wants to author AI without programming. In particular, the LfD technique presented in this chapter has been used to power the social gaming website *Make ME Play ME*, in which users compete to create good AIs by demonstration.

The work presented in this chapter indicates that to enable user-generated AI, we need to address both technical and user-interface problems. D2 and Meta-Darmok are attempts at addressing the technical challenges, and MMPM and the iIDE are attempts at addressing the user-interface problems.

Moreover, although the techniques presented in this chapter are useful for achieving user-generated AI, we have listed a number of open problems that need to be solved before they can be applied to a broad variety of games by end-users. In the case of LfD, the two main open problems of the approaches presented here are how to present the learned strategies to the user in a human-understandable way, and how to achieve generic and efficient plan adaptation (for adapting learned plans to new situations). Concerning self-adaptation, the main problems of a failure-pattern-based approach are enabling the easy definition of failure patterns for end-users in an intuitive way.

References

1. Aamodt, A., Plaza, E.: Case-based reasoning: Foundational issues, methodological variations, and system approaches. Artificial Intelligence Communications **7**(1), 39–59 (1994)
2. Aha, D. (ed.): Lazy Learning. Kluwer Academic Publishers, Norwell, MA, USA (1997)
3. Aha, D.W., Molineaux, M., Ponsen, M.J.V.: Learning to win: Case-based plan selection in a real-time strategy game. In: ICCBR, pp. 5–20 (2005)
4. Bauer, M.A.: Programming by examples. Artificial Intelligence **12**(1), 1–21 (1979)
5. Berger, L.: Scripting: Overview and code generation. In: S. Rabin (ed.) AI Game Programing Wisdom, pp. 505–510. Charles River Media (2002)
6. Cox, M.T., Ram, A.: Introspective multistrategy learning: on the construction of learning strategies. Artificial Intelligence **112**(1-2), 1–55 (1999)
7. Fikes, R., Nilsson, N.J.: Strips: A new approach to the application of theorem proving to problem solving. Artificial Intelligence **2**(3/4), 189–208 (1971)

8. Floyd, M.W., Esfandiari, B., Lam, K.: A case-based reasoning approach to imitating robocup players. In: Proceedings of the Twenty-First International Florida Artificial Intelligence Research Society (FLAIRS), pp. 251–256 (2008)
9. Kolodner, J.: Case-based reasoning. Morgan Kaufmann (1993)
10. Lieberman, H.: Tinker: a programming by demonstration system for beginning programmers. In: Watch what I do: programming by demonstration, pp. 49–64. MIT Press, Cambridge, MA, USA (1993)
11. Lozano-Pérez, T.: Robot programming. In: Proceedings of IEEE, vol. 71, pp. 821–841 (1983)
12. Mehta, M., Ontañón, S., Ram, A.: Using meta-reasoning to improve the performance of case-based planning. In: Case-Based Reasoning, ICCBR-2009, no. 5650 in Lecture Notes in Artificial Intelligence, pp. 210–224. Springer-Verlag (2009)
13. Murata, T.: Petri nets: Properties, analysis and applications. Proceedings of the IEEE **77**(4), 541–580 (1989)
14. Nakanish, J., Morimoto, J., Endo, G., Cheng, G., Schaal, S., Kawato, M.: Learning from demonstration and adaptation of biped locomotion with dynamical movement primitives. In: Workshop on Robot Learning by Demonstration, IEEE International Conference Intelligent Robots and Systems (2003)
15. Nicolescu, M.N.: A framework for learning from demonstration, generalization and practice in human-robot domains. Ph.D. thesis, University of Southern California, Los Angeles, CA, USA (2003). Adviser-Maja J. Mataric
16. Ontañón, S., Bonnette, K., Mahindrakar, P., Gómez-Martín, M., Long, K., Radhakrishnan, J., Shah, R., Ram, A.: Learning from human demonstrations for real-time case-based planning. IJCAI-09 Workshop on Learning Structural Knowledge From Observations (STRUCK-09) (2009)
17. Ontañón, S., Mishra, K., Sugandh, N., Ram, A.: On-line case-based planning. Computational Intelligence Journal **26**(1), 84–119 (2010)
18. Puga, G.F., Gómez-Martín, M.A., Díaz-Agudo, B., González-Calero, P.A.: Dynamic expansion of behaviour trees. In: AIIDE (2008)
19. Ryan Houlette, D.F.: The ultimate guide to fsms in games. In: S. Rabin (ed.) AI Game Programing Wisdom 2, pp. 283–302. Charles River Media (2003)
20. Schaal, S.: Learning from demonstration. In: Advances in neural information processing systems 9, pp. 1040–1046. MIT press (1997)
21. Spalazzi, L.: A survey on case-based planning. Artificial Intelligence Review **16**(1), 3–36 (2001)
22. Spronck, P., Kuyper, S.I., Postma, E.: Difficulty scaling of game AI. In: Proceedings of the 5th International Conference on Intelligent Games and Simulation (GAME-ON 2004), pp. 33–37 (2004)
23. Spronck, P., Ponsen, M., Sprinkhuizen-Kuyper, I., Postma, E.: Adaptive game AI2 with dynamic scripting. Machine Learning **63**(3), 217–248 (2006)
24. Sugandh, N., Ontañón, S., Ram, A.: On-line case-based plan adaptation for real-time strategy games. In: AAAI 2008, pp. 702–707 (2008)
25. Virmani, S., Kanetkar, Y., Mehta, M., Ontañón, S., Ram, A.: An intelligent IDE for behavior authoring in real-time strategy games. In: AIIDE (2008)
26. Zang, P., Mehta, M., Mateas, M., Ram, A.: Towards runtime behavior adaptation for embodied characters. In: IJCAI'07: Proceedings of the 20th international joint conference on Artificial intelligence, pp. 1557–1562. Morgan Kaufmann Publishers Inc., San Francisco, CA, USA (2007)

Game AI as Storytelling

Mark Riedl, David Thue, and Vadim Bulitko

Abstract Much research on artificial intelligence in games has been devoted to creating opponents that play competently against human players. We argue that the traditional goal of AI in games-to win the game-is but one of several interesting goals to pursue. We promote the alternative goal of making the human player's play experience "better," i.e., AI systems in games should reason about how to deliver the best possible experience within the context of the game. The key insight we offer is that approaching AI reasoning for games as "storytelling reasoning" makes this goal much more attainable. We present a framework for creating interactive narratives for entertainment purposes based on a type of agent called an experience manager. An experience manager is an intelligent computer agent that manipulates a virtual world to dynamically adapt the narrative content the player experiences, based on his or her actions and inferences about his or her preferred style of play. Following a theoretical perspective on game AI as a form of storytelling, we discuss the implications of such a perspective in the context of several AI technological approaches.

1 From Adversarial Agents to Experience Management

Historically, games have been played between human opponents. However, with the advent of the computer came the notion that one might play with or against a computational surrogate. Dating back to the 1950s with early efforts in computer chess, approaches to game artificial intelligence (AI) have been designed around adversarial, or zero-sum, games. The goal of intelligent game-playing agents in these cases is to maximize their payoff. Simply put, they are designed to win the game. Central to the vast majority of techniques in AI is the notion of optimality, implying that the best performing techniques seek to find the solution to a problem that will

M. Riedl (✉)
Georgia Institute of Technology, Atlanta, GA 30332, USA
e-mail: riedl@cc.gatech.edu

P.A. González-Calero and M.A. Gómez-Martín (eds.), *Artificial Intelligence for Computer Games*, DOI 10.1007/978-1-4419-8188-2_6,

result in the highest (or lowest) possible evaluation of some mathematical function. In adversarial games, this function typically evaluates to symmetric values such as $+1$ when the game is won and -1 when the game is lost. That is, winning or losing the game is an outcome or an end. While there may be a long sequence of actions that actually determine who wins or loses the game, for all intents and purposes, it is a single, terminal event that is evaluated and "maximized." In recent years, similar approaches have been applied to newer game genres: real-time strategy, first person shooters, role-playing games, and other games in which the player is immersed in a virtual world. Despite the relative complexities of these environments compared to chess, the fundamental goals of the AI agents remain the same: to win the game.

There is another perspective on game AI often advocated by developers of modern games: AI is a tool for increasing engagement and enjoyability. With this perspective in mind, game developers often take steps to "dumb down" the AI game playing agents by limiting their computational resources [30] or making suboptimal moves [58] such as holding back an attack until the player is ready or "rubber banding" to force strategic drawbacks if the AI ever gets the upper hand. The game-playing agent is adversarial but is designed to be noncompetitive through the use of ad hoc rules with the intention that the player feel powerful.

In this chapter, we focus on game AI which, instead of being designed to win more often, reasons in a *principled* manner about how to make the human player's experience in a game or virtual world more enjoyable. While the outcome of a game is important, it is not the only aspect of a game that a player evaluates. *How* one reaches the ending can often be just as, if not more, important than what the ending is; a hard fought battle that results in a loss can be more enjoyable than an easy win. Extrapolating from the observation that experience can be more important than outcome, we suggest that the goal of computer game AI is to reason about and deliver an enjoyable experience. Game AI thus becomes a tool in the arsenal of the game designer, to be used whenever one would want a real person to play a given role but no one is available or willing. Examples of such roles are:

- Opponents, companions, and NPCs that play roles that are not "fun" to play such as shopkeepers, farmers, and victims
- Dungeon master
- Plot writer
- Game designer

As we go down this list, game AI is charged with taking progressively more responsibility for the quality of the human player's experience in the game. To leverage this model, we redefine the task of game AI agents as the creation of an enjoyable player experience, and define payoffs that allow them to optimize the particular qualities of the experience that its designers might desire. Regardless of whether the AI agent is choosing how to oppose or assist the player[1] or how the storyline should unfold, the player's enjoyment is its central concern.

[1] Roberts et al. [48] use the term *beyond adversarial* to denote that a game AI system can choose to help or hinder the player based on its assessment of the player's past, current, and future experience.

1.1 *Reasoning About Experience as Proxy for Designer and Player*

We define an experience as one or more interrelated events directly observed or participated in by a player. In games, these events are causally linked series of challenges that play out in a simulated environment [49]. Intuitively, it is the job of the game designer to make decisions about how to shape the player's experience in a virtual world to make it enjoyable. One way game designers do this is using a "story on rails" to lead players through a dramatically engaging sequence of challenges. While game design approaches have been effective in creating engaging and enjoyable experiences, there is a growing trend toward greater player agency and greater content customization, and neither of these can be achieved easily at the time of game design:

- *Player agency.* Player agency is the ability for the player to do whatever he or she wants at any time. While player agency is typically very high at the action level – the player has the ability to move about the environment and perform actions – agency at the plot level has typically been restricted to a single storyline or a small number of storyline branches. One reason for this restriction is the combinatorial explosion of authoring storyline branches [9,46]; the amount of content that must be authored at least doubles at every branching point, yet a player will only see one branch.
- *Customization.* Players enjoy having opportunities to experience game play that is consistent with their preferred play style [52]. In one study, Thue et al. [53] demonstrated that the player model-based adaptation of players' in-game experiences resulted in greater reports of fun. The information required to make customization decisions, however, is not available at design time; it must be learned by observing the player during game play.

In short, to achieve greater levels of player agency and greater levels of content customization, the computational game system must assume responsibility for the player's experience during play time. The role of determining what the player's experience should be (including how the game world responds to the player's actions) can be delegated to the computational game system itself.

1.2 *Leveraging Storytelling*

How can an intelligent system in a game or virtual world make decisions about, or indirectly influence, the events that occur in the simulated environment such that the positive qualities of the player's experience are increased? As with game designers, an intelligent system can leverage correlations between experience and narrative to reason about how to manage a player's experience. A narrative is *the recounting of a sequence of events that have a continuant subject and constitute a whole* [40].

Thus, both "experience" and "narrative" are descriptions of sequences of events. From a game design perspective, an experience is a description – at some level of abstraction or specificity – of events that are *expected* to unfold.

An *Experience Manager* – a generalization of the concept of *Drama Manager* first proposed by Laurel [26] and first investigated by Bates and colleagues (cf. [5, 25]) – is an intelligent system that attempts to coerce the state of the world such that a structured narrative unfolds over time without reducing the perceived agency of the interactive player. An Experience Manager uses the principle of narrative to *look ahead* into possible futures of the player's experience to determine what *should* happen in the world over time to bring about an enjoyable, structured experience. The projection of a narrative sequence into the future enables the Experience Manager to evaluate the global structure of possible player experiences in a way that cannot be achieved by looking at any single world state in isolation. The question that must be addressed is: in light of an interactive player, how can a computational system project a narrative into the future toward maximizing the positive qualities of the player's experience?

In artificial intelligence, problems are often modeled as state spaces, where every point in this abstract space is a particular configuration of the game environment. While a game playing agent may attempt to maximize expected payoff by choosing which state to transition to next, a system attempting to optimize player experience must choose a *sequence* of states through which the game should transition. We refer to a sequence of state transitions as a *trajectory* through state space. Choosing a trajectory is nontrivial; even when the state space is finite, the number of possible trajectories can be infinite when loops are allowed, as is often the case with stories. Figure 1a, b shows two possible trajectories through state space. Every point in the oval is a possible state configuration for the entire virtual world. Actions, performed by the player or the computational system (possibly through the actions of non-player characters), cause transitions from one world state configuration to the next; in any given state, there are potentially many actions that can be performed. If we assume that states of the right side of the oval are terminal states, then the trajectory in Fig. 1a may be one that reaches a terminal state in the fewest transitions, a metric classically used to determine the efficiency and optimality of a solution.

The trajectory in Fig. 1b, however, may enter parts of state space that, when taken together, are more interesting, dramatic, or pedagogically meaningful. The challenge is to computationally find the one trajectory in the space of all trajectories that will optimize the player's experience (Fig. 1a–c).

Because a narrative is the recounting of a nonrandom sequence of events, *any trajectory through a state space is a narrative*. Making the connection between a trajectory of states in a game and a narrative enables us to cast the search for a particular trajectory as the problem of generating a story. The computational generation of stories is still an open problem; story generation algorithms exist that can produce event sequences for games or virtual worlds, although not at a level comparable to human creative performance. Further, the problem of story generation is made more complex by the fact that an interactive player has the ability to act in the world, making it impossible to guarantee that any narrative sequence will unfold as expected. That is, if the player, knowingly or inadvertently, performs

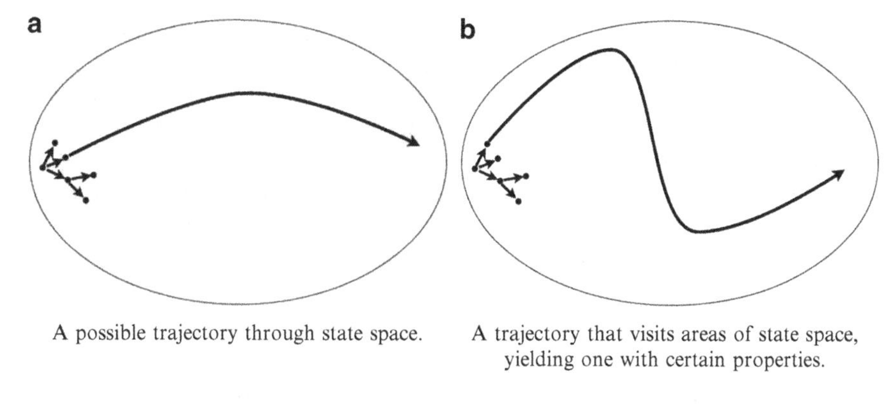

A possible trajectory through state space.

A trajectory that visits areas of state space, yielding one with certain properties.

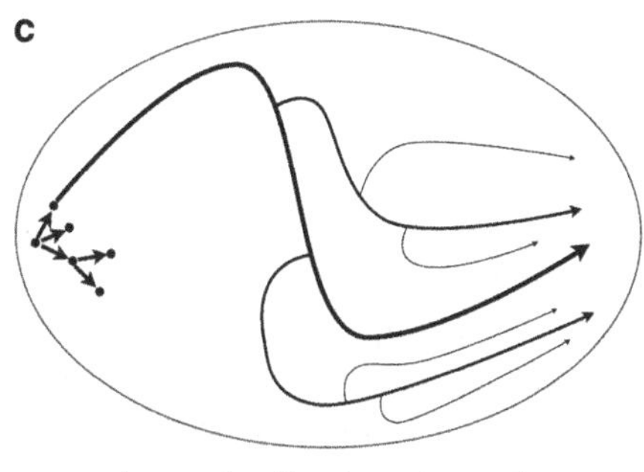

Accounting for player interaction.

Fig. 1 The experience management problem is to compute trajectories through state space

actions that cause the virtual world state to deviate from the expected trajectory, the next best trajectory must be computed, as shown in Fig. 1c. In that sense, managing the player's experience is the problem of searching for many alternative stories. In this chapter, we describe a technique for experience management that uses story generation technologies to manage an interactive player in a virtual world. The goal of the approach is to coerce the events in a virtual world such that the player has an enjoyable experience with certain well-defined narrative properties.

The heart of experience management is the tension between meaningful player agency and the desire to bring about a narrative experience that is coherent and conforms to the designer's pragmatic and esthetic ideals [43]. Having considered the Experience Manager as a proxy for the designer, we may also consider the extent to which the player's preferences are part of the definition of the "optimal" narrative trajectories. A player model informs the trajectory search about what the player will find interesting and enjoyable, such that when there are multiple ways of achieving the designer's pragmatic and esthetic requirements, the trajectory that appeals most to the player can be selected.

In this chapter, we discuss how game AI can be reinterpreted as storytelling for the purpose of reasoning about the human player's experience, thereby creating greater player agency through more meaningful interactions, and affording more customization of experience. We begin with an overview of *narrative intelligence* – the ability to reason about narrative – and narratological foundations of computational storytelling. In Sect. 3, we provide a computational representation of narrative and narrative construction. In Sect. 4, we describe how an Experience Manager can use the ability to computationally generate narrative to manage the player's interactive experience in a game or virtual world. Finally, in Sect. 5, we address customization of the player's interactive experience through learning a model of the player and using it to optimize his or her particular narrative trajectory.

2 Narrative Intelligence and Narratological Foundations

Narrative as entertainment, in the form of oral, written, or visual storytelling, plays a central role in many forms of entertainment media, including novels, movies, television, and theater. Narrative is also used in education and training contexts to motivate and to illustrate. One of the reasons for the prevalence of storytelling in human culture may be due to the way in which narrative is a cognitive tool for situated understanding [10, 18, 19, 21, 36]. There is evidence that suggests that we, as humans, build cognitive structures that represent the real events in our lives using models similar to the ones used for narrative to better understand the world around us [10]. Our understanding of the world is achieved by "constructing reality" as a sequence of related events from our senses [11]. While we tend to understand inanimate objects through cause and effect, we attempt to understand the intentional behavior of others through a sophisticated process of interpretation with narrative at its core [10]. This *narrative intelligence* [7, 33] is central in the cognitive processes that we use across a range of experiences, from entertainment to active learning.

Narratologists break narrative down into at least two layers of interpretation: *fabula* and *sjuzet* [3]. The *fabula* of a narrative is an enumeration of all the events that occur in the story world between the time the story begins and the time the story ends. The events in the *fabula* are temporally sequenced in the order that they occur, which is not necessarily the same order in which they are told. The *sjuzet* of a narrative is a subset of the *fabula* that is presented via narration to the audience. If the narrative is written or spoken word, the narration is in natural language. If the narrative is a cinematic presentation, computer game, or virtual world, the narration is through the actions of characters and the camera shots that capture that action. While it is the narrated *sjuzet* that is directly exposed to the audience, it is the *fabula* of a narrative that is the content of the narrative (i.e., what the narrative is about).

In this chapter we focus on *fabula*: what happens (or what is expected to happen) in the virtual world or game. Readers interested in how stories can be computationally structured at the *sjuzet* level should see Montfort [37], Cheong and Young [14], Bae and Young [2], and Jhala [24].

There are many aspects that determine whether a story is accepted by the audience as "good." Many of these aspects are subjective in nature, such as the degree to which the audience empathizes with the protagonist. Other aspects appear to be more universal across a wide variety of genres. Cognitive psychologists have determined that the ability of an audience to comprehend a narrative is strongly correlated with the causal structure of the story [8, 20, 21, 56] and the attribution of intentions to the characters that are participants in the events [19–21]. Story comprehension requires the perception of causal connectedness of story events and the ability to infer the intentions of characters. Accordingly, we assert that two nearly universal qualities of narratives are *logical causal progression* and *character believability*.

The causality of events is an inherent property of narratives and ensures a whole and continuant subject [13]. Causality refers to the notion that there is a relationship between temporally ordered events such that one event changes the story world in a particular way that enables future events to occur [54]. For a story to be considered successful, it must contain a degree of causal coherence that allows the audience to follow the logical succession of events and predict possible outcomes. One can think of the property of logical causal progression as the enforcement of the "physics" of the story world in the sense that there are certain things that can and cannot happen based on the actual state of the story world and the characters within it. For example, in fairy tales, the world is such that wild animals such as wolves can eat people without killing them. Character believability [6] is the audience perception that arises when the actions performed by characters do not negatively impact the audience's suspension of disbelief. Goal-oriented behavior is a primary requirement for believability [12,31]. Specifically, we, as humans, ascribe intentionality to agents with minds [16]. The implication is that if a character is to be perceived as believable, one should be able to, through observations of the character, infer and predict its motivations and intentions. For a greater analysis of goal-directed behavior in character believability, see Riedl and Young [47].

3 Computing Narrative Structure

Addressing experience management as story generation, there are two problems to consider. The first is how to computationally model narrative structure. The second is how to computationally model the process of constructing narrative and managing interactive experiences. This section addresses the computational representation of narrative in detail, but only touches on algorithms for generation due to the fact that there are still many open research problems that remain to be addressed. The general consensus among psychologists and computer scientists is that a narrative can be modeled as a semantic network of concepts [20, 51, 55, 60]. Nearly all cognitive representations of narrative rely on causal connections between story events as one of the primary elements that predict human narrative comprehension. Following others [28, 39, 43, 45, 47, 60], we use AI plan-like representations of narrative as transitions through state space. Plan representations have been used in numerous narrative intelligence systems; they correlate well with the narratological and cognitive constructs that have been associated with narrative reasoning.

3.1 Narrative as Plans

Partial-order causal link (POCL) plans [57], in particular, have been used successfully to computationally reason about narrative structure because of strong correlations between representation and cognitive and narratological concepts [15, 60]. A POCL plan is a directed acyclic graph in which nodes are operations (also called actions) which, when executed, change the world state. Arcs capture causal and temporal relations between actions. A *causal link*, denoted $a_i \rightarrow^c a_j$, captures the fact that the execution of action a_i will cause condition c to be in true in the world, and that condition is necessary for the execution of subsequent action a_j. Causal links, unique to POCL plans, are representationally significant due to the importance of causality in narratives. *Temporal links* capture ordering constraints between actions when one action must be performed before another. Temporal and causal links create a partial ordering between actions, meaning that it is possible that some actions can occur during overlapping time intervals.

A narrative is a sequence of events – significant changes to the state of the story world. The mapping of narrative to plan is straightforward. Events are represented by plan actions, which are partially ordered with respect to each other by the temporal links. The term "event" captures the nuance that not all changes to the world state are intentional on behalf of some agent or character. Thus, some events can be accidents, automatic reactions to other changes, and forces of nature. Partial ordering is a favorable feature of a story representation because it is often the case that actions in the *fabula* occur simultaneously. In the remainder of this chapter, we will use the terms "action" and "event" interchangeably. Note that a narrative plan for an interactive game or virtual world contains events to be initiated by the player and non-player characters. Figure 2 shows an example of a plan representing a narrative sequence for a simplified version of *Little Red Riding Hood* [22]. Boxes represent events. Solid arrows are causal links where the labels on the links describe the relevant conditions. Dashed arrows represent temporal constraints between events. For clarity not all causal and temporal links are shown. There are three types of special constructs shown in the figure. The *initial state* is a description of the story world as a set of logical propositions. The initial state specifies characters in the story world, the properties of characters, and relationships between characters, props, and the world. The *outcome* is a description of how the story world should be different after the story completes. In Fig. 2, the outcome is that the Granny character has cake, Granny is not in the state of being eaten, and Little Red Riding Hood ("Red" for short) is also not in the state of being eaten. Finally, *author goals* are intermediate states that must be achieved as some point during the course of the story. In the example, the two author goals are that Granny becomes eaten by something and Red becomes eaten by something. Author goals are used to preserve the authorial intent of the designer, as described in Sect. 3.2.

Next, we overview how plans are computationally constructed. Planners are search algorithms that solve the planning problem: given an initial world state, a domain theory, and a goal situation, find a sound sequence of operators that transforms the world from the initial state into a state in which the goal situation holds.

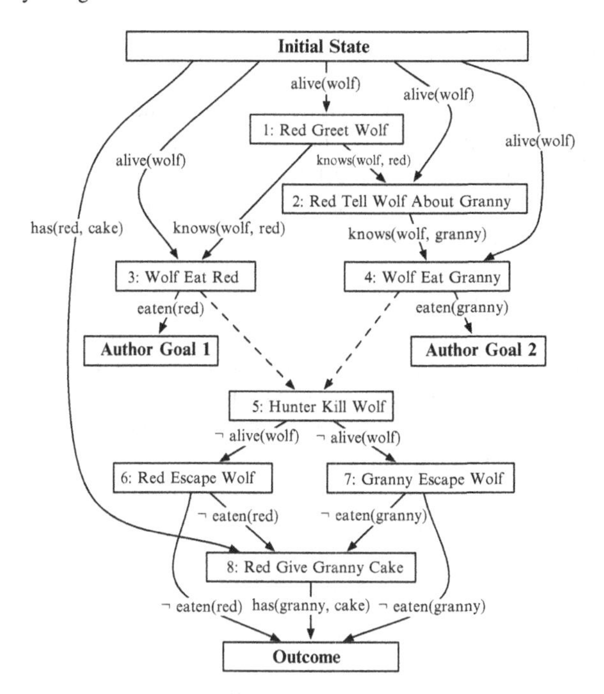

Fig. 2 The story of Little Red Riding Hood represented as a partially ordered plan

---

Action: Eat (?wolf, ?victim)

```
Precondition: wolf(?wolf), person(?victim), alive(?wolf), alive(?victim),
              ¬eaten(?wolf), ¬eaten(?victim)
Effect: eaten(?victim), in(?victim, ?wolf), full(?wolf)
```

Action: Tell-About (?speaker, ?hearer, ?topic)

```
Precondition: character(?speaker), character(?hearer), alive(?speaker),
              alive(?hearer), ¬eaten(?speaker), ¬eaten(?hearer),
              knows(?speaker, ?hearer), knows(?speaker, ?topic),
              ?speaker ≠ ?hearer
Effect: knows(?hearer, ?topic)
```

Fig. 3 Portion of a domain library for Little Red Riding Hood

The *domain theory* describes the "physics" of the world – how the world works and how it can be changed. The domain theory is often a library of action (or event) templates where applicability criteria and world state update rules are specified through logical statements. Specifically, events have *preconditions* and *effects*. The precondition of an event is a logical statement that must be true in the world for the event to legally occur. The effect of an event is a logical statement that describes how the world would be different if the events were to occur. Figure 3 shows two event templates from the domain library for the Little Red Riding Hood world.

There are many algorithms that solve the planning problem. In this section, we highlight *partial-order planning* (POP) [38,57]. POP planners are refinement search algorithms, meaning they inspect a plan, identify a flaw – a reason why the current plan being inspected cannot be an actual solution – and attempt to revise the plan to eliminate the flaw. The process iteratively repairs one flaw at a time until no flaws remain. It is often the case that there is more than one way to repair a flaw, in which case the planner picks the most promising repair, but remembers the other possibilities. Should it make a mistake, the planner can *backtrack* to revisit any previous decision point. We favor POP because the particular way in which flaws are identified and revised in POP is analogous to cognitive planning behavior in adult humans when faced with unfamiliar situations [41].

The refinement search process starts with an empty plan. A flaw is detected, and zero or more new plans are produced in which the flaw is repaired (and often introducing new flaws). These plans become part of the fringe of a space consisting of all possible complete and incomplete plans. The process is repeated by picking the most promising plan on the fringe and iterating.

In POP, there are two types of flaws: *open conditions* and *causal threats*. An open condition flaw exists when an event (or the outcome) has a precondition that has not been satisfied by the effect of a preceding event (or the initial state). An open condition flaw is repaired by applying one of the following strategies:

1. Select an existing event in the plan that has an effect that unifies with the precondition in question.
2. Select and instantiate an event template from the domain library that has an effect that unifies with the precondition in question.

A causal threat flaw occurs when the temporal constraints do not preclude the possibility that an action a_k with effect $\neg c$ can occur between a_i and a_j when there is a causal link $a_i \rightarrow^c a_j$ requiring that c remain true. Causal threats are repaired by adding additional temporal constraints that force a_k to occur before a_i or after a_j. By iteratively repairing flaws, the current plan progressively gets closer to a solution. The algorithm terminates when it finds a plan that has no flaws. More details on POP are provided by Weld [57].

3.2 Preserving Designer Intent

A narrative generator assumes responsibility for the structure of the player's experience during gameplay. It is, however, desirable that a designer is able to constrain the space of possible experiences the player can have to enforce a particular esthetic or pragmatic vision. We extend the standard POCL plan representation to include *author goals* [42], partially specified intermediate states that the story must pass through at some point before the story concludes. Potential solutions that do not satisfy each author goal state description at some point between the initial state and the end state are pruned from consideration. Author goals serve two important purposes. First, author goals constrain the narrative search space such that it is impossible for a generator to produce a story that does not meet certain criteria imposed by the designer.[2] Second, author goals can be used to force complexity in the story. The importance of author goals as part of the narrative representation becomes clear in the context of the Little Red Riding Hood example. Without author goals, achieving the outcome – Granny has cake, Granny is not eaten, and Red is not eaten – is trivial. Red need only give some cake to Granny, which can be achieved with a single event. The author goals – Granny is eaten, and Red is eaten – force the story generator to figure out how to have both Granny and Red eaten and then later saved.

3.3 Generating Believable Stories

If plans are good representations of narratives, might it also make sense to use planning algorithms to construct narratives? Young and Saver [59] provide neurological evidence of functional similarity between planning and narrative generation in the human brain. Planning algorithms, however, are general problem solvers that make strong assumptions about the nature of the problem being solved. Specifically, a planner is an algorithm that attempts to find a sequence of operations that transforms the world state from an initial configuration into one in which a given goal situation holds. While a resultant set of operations – a plan – can be considered a narrative, that narrative is unlikely to be believable or to contain esthetic features such as a dramatic arc that would be favorable for the task of creating engaging experiences [47]. The reason that conventional planners are not guaranteed to generate believable narrative plans is because of their emphasis on achieving valid plans; they disregard the requirement that characters will appear motivated by intentions other than the author's goals. Even with a heuristic that favors plans in which characters appear believable, it is possible for a conventional planner to return a plan that is not believable when it finds a shorter, valid solution before it finds a longer, valid, and believable solution.

To reliably generate narrative plans in which characters appear believable, narrative planners must use new definitions for plan completeness that include believability, coupled with mechanisms for selecting actions that move the planner toward complete, believable solutions. Extensions to POP, implemented in the FABULIST story generation system [47], allow planners to search for narrative sequences that are both logically and causally coherent but also present events that explain the underlying motivations of characters. This is one step toward computationally achieving the "illusion of life" necessary for suspension of disbelief [6]. Future work in story generation must also consider esthetics such as dramatic arc – the cadence of rising action, climax, and falling action – and suspense. While efforts are underway to explore computational reasoning about such story esthetics (cf. [1,17]), there

[2] In the absence of a well-defined evaluation function that can rate the "goodness" of a narrative trajectory, the designer's intent is the only guidance the story generator has.

are many open research questions to be addressed in the pursuit of computational systems that can assume full responsibility for the quality of a player's interactive experience. In the remainder of this chapter, we will describe our approach to experience management in the context of simple POP, although more sophisticated algorithms exist that are keyed to the specific problem of generating believable narrative sequences.

4 Experience Management

Player experience in a virtual world or game can be expressed as a narrative, projecting an ideal trajectory of state transitions into the future. This narrative is *not* necessarily the sequence of moves that a rational computer opponent would take to maximize expected payoff but rather the one that delivers a "good" experience to the player. In a virtual world modeled after Little Red Riding Hood, this may be the sequence that raises the stakes for the player but then allows the player to overcome adversity to save the day. In a game of chess, this may be the sequence that sets up a dramatic come-from-behind victory. Thus far, however, we have not addressed the fact that the player is not just another character in the story, but a human with his or her own goals and the ability to make gameplay choices that differ from the idealized narrative sequence. That is, the human player neither knows the script nor is expected to follow it.

Experience management is the process whereby a player's agency is balanced against the desire to bring about a coherent, structured narrative experience. On one hand, we want the player to have the perception that he or she has the ability to make decisions that impact the world in a meaningful way (e.g., at the plot level). On the other hand, the designer wants the player to have an experience that meets certain esthetic and pragmatic guidelines. Can we allow meaningful player agency while still achieving the goal of bringing about an experience that has the features desired by the designer? Designers of heavily plot-driven computer games often resort to a "story on rails" approach, where although there may be an *appearance* of agency, the world is structured so that a single pre-scripted plot sequence unfolds; side-quests are then often added to enhance the player's feeling of agency. The "story on rails" approach is diametrically opposite of simulation style games, in which there is no pre-scripted plot sequence, and any narrative structure emerges from the interactions of autonomous non-player characters and human players.

Our approach to experience management, as implemented in the AUTOMATED STORY DIRECTOR framework [44] and the MIMESIS system [62], balances player agency and narrative structure by allowing meaningful player agency and then *generating* novel narrative trajectories when the player, intentionally or inadvertently, exerts their agency. Consider trajectory space – the set of all possible trajectories through state space. One trajectory, the *exemplar trajectory*, is the human designer's preferred story; it is the best possible experience according to that designer. The exemplar trajectory projects the player's actions and non-player character actions into the future. Players may still exert their agency, however, and we categorize their actions as follows [43]:

- *Constituent* – the player knowingly or unknowingly performs the action that is listed as the next action in the narrative. For example, after the Wolf has eaten Red and Granny, the player, in the role of the hunter, kills the Wolf.
- *Consistent* – the player performs an action that is not part of the narrative but does not significantly alter the state of the world and the narrative sequence can continue. For example, early in the game, the player talks to Red.
- *Exceptional* – the player performs an action that is not part of the narrative, and the world state is changed such that some portion of the narrative cannot continue. For example, the player kills the Wolf before the Wolf meets Red, or the player takes the cake away from Red.

In the case that the player performs an exceptional action, the Experience Manager must figure out how to allow the player's action[3] and still achieve an experience with the requisite structure. Note that the exceptional player action may not immediately threaten the narrative, as the change to the world may impact an action that is projected to occur far downstream in time. Handling an exceptional player action is tantamount to finding the next best trajectory, given a world state altered by the player's exceptional action.

4.1 Anticipating Necessary Narrative Plan Adaptations

Using plan structures to model narrative is advantageous because, by capturing the causal relationships between actions, a narrative plan can be analyzed for points in which exceptional player actions are possible. That is, assuming the narrative plan executes as expected, we can look into the future and identify possible exceptional actions. We use a technique similar to that described by Riedl et al. [43] to analyze the causal structure of the scenario to determine all possible inconsistencies between plan and virtual world state that can occur during the entire duration of the narrative. Inconsistencies arise due to exceptional player actions performed in the world. The technique identifies intervals of the narrative plan during which it is possible for an exceptional action to occur.

For every possible inconsistency that can arise that threatens a causal link in the plan, an alternative narrative plan is generated. For each possible inconsistency that

[3] One can also consider attempting to prevent the exceptional action in a natural and unobtrusive manner. Riedl et al. [43] describe a technique called *intervention* whereby the exceptional action is surreptitiously replaced by a nearly identical action that does not affect the world state in a way that prevents the narrative from progressing. For example, if shooting a character prevents that character from performing a critical task in the future, then *shoot* can be replaced by *gun-jam* that prevents the character from dying and allowing the narrative to continue. However, if intervention is chosen, it effectively removes player agency, which may or may not be noticed by the player.

can arise, we use the following repair process to find an alternative trajectory. First, we assume that the narrative will progress as expected until the threatened interval begins. Next, we assume the worst case: that the player will perform the exceptional action that creates the inconsistency. By simulating the execution of the exceptional player action, we can infer the state that the world would be in if the action were to occur. Finally, the following repair processes are tried in order until one succeeds in generating a narrative that meets the designer's intent:

(a) The threatened causal link is removed, leaving an open condition flaw on the terminus event, and the planner is invoked.
(b) The threatened causal link is removed, the terminus event and all other events (except author goals) that are causally downstream (e.g. there is a path from the threatened causal link to a given event through the directed graph of causal links) are removed, open condition flaws are identified on the remaining events, and the planner is invoked.
(c) The threatened causal link is removed, the terminus event and all other events (including author goals) that are causally downstream are removed, open conditions flaws are identified on the remaining steps, and the planner is invoked.
(d) The remaining plan is discarded, a new outcome situation and new author goals are selected, and the planner is invoked.

To illustrate the tiers of repair strategies, consider the narrative plan in Fig. 4a. Event e_1 establishes condition c in the world, which is necessary for event e_2. Suppose it is possible for the player to perform an action that causes $\neg c$ to become true during the interval between the completion of e_1 and the beginning of e_2. The possible inconsistency is found during causal analysis, and the tier (i) strategy is invoked. A copy of the plan is made and updated to reflect the state the world would be in should all events preceding the interval in question have occurred. That is, the initial

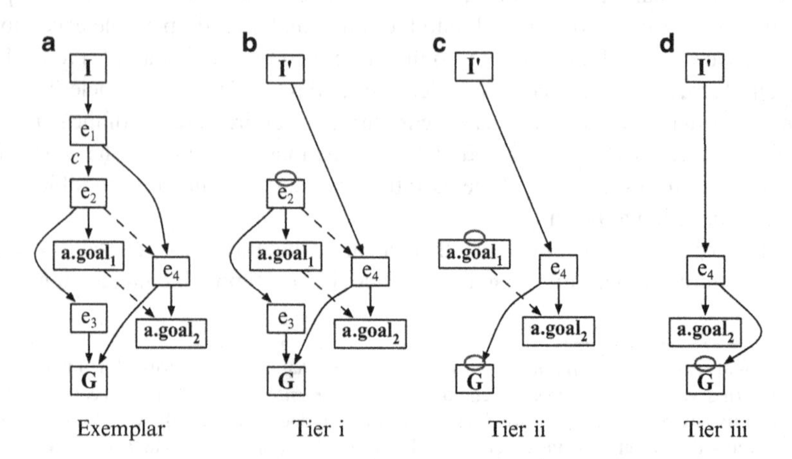

Fig. 4 Illustration of the tiered re-planning strategies considering a single possible inconsistency resulting in $\neg c$ in exemplar (**a**). (**b**)–(**d**) show how the exemplar is prepared for re-planning for each tier

state now represents the world state after e_1 has occurred, and action e_1 is no longer part of the narrative. The tier (i) strategy removes only the causal links in the interval in question that are threatened by the exceptional player action. Figure 4b shows the copy of the plan after tier (i) preprocessing but before the story generator is invoked to fill back in causally necessary events. Ovals indicate flaws in the plan due to preprocessing. Replanning will most likely result in the insertion of new events before e_2 that reestablish c in the world.

Suppose that the tier (i) strategy fails, the story generator cannot find any partially ordered sequence of new events that can fill the gap created by removing the threatened causal link. The Experience Manager advances to the tier (ii) strategy, and removes threatened causal links, the events satisfied by the threatened links, and all events that are causally downstream except author goals. A causally downstream event is any action e_i such that there is a path in the graph of causal links from a removed event to e_i. In this example, action e_3 is causally downstream but e_4 is not. Figure 4c shows the copy of the plan after tier (ii) preprocessing. The tier (iii) strategy is similar to tier (ii) except that causally downstream author goals are also removed. The underlying assumption is that tier (ii) failed because the author goals were interfering with the ability of the story generator to find a valid plan. Figure 4d shows the copy of the plan after tier (iii) preprocessing. Finally, should all other strategies fail, the tier (iv) strategy (not shown) deletes all actions in the plan, replaces the outcome situation G with a new outcome situation G', and instantiates any number of new author goals. The new outcome and author goals come from a list of alternative author goals specified at design time by the human designer. If the final tier of replanning fails, we resort to a non-managed virtual world, relying completely on game play dynamics and the autonomy of non-player characters to create an emergent narrative experience.

We use the tiered strategy approach to compensate for the fact that story planners are not yet at human-level ability for story creation. In the absence of a story planner that can reliably evaluate the "goodness" of a narrative sequence, the tiered strategy approach is built on the assumption that the human-authored exemplar narrative is the ideal experience, and that any necessary changes should preserve the original narrative structure as much as possible. This assumption is not true in all cases and can result in situations where the Experience Manager attempts to undo the consequences of the player's actions [44]. As story generation techniques improve (see Sect. 3.3 for pointers to potential improvements), reliance on such assumptions will become unnecessary, simplifying the operation of the Experience Manager.

4.2 *Computation of Contingencies*

Story replanning is performed offline to avoid delays due to computation [43, 44]; for any sufficiently rich world, the online generation of narrative structure can exceed acceptable response times in an interactive game or virtual world. The result of the offline replanning process is a tree of contingency plans in which each plan represents a complete narrative starting either at the initial world state (for the exemplar) or at the point in which an inconsistency can occur at play time. If the player performs an exceptional action, the system simply looks up the appropriate branch in the tree of contingencies and seamlessly begins using the new trajectory to manage the player's experience from that point on. The contingency tree is necessary for dynamic execution; by pre-generating the tree, an Experience Manager can rapidly switch to alternative narrative plans when player actions make this necessary. Riedl and Young [46] show that the contingency tree is functionally equivalent, but more expressive, than a choose-your-own-adventure style branching story. The additional expressivity comes from the fact that player actions can be performed at any time (e.g., in any interval). Note that a tree of contingency plans can be potentially infinite in depth. We use a simple user model to determine which exceptional actions are most probable, and focus on making those parts of the tree more complete (cf. [44]). Additionally, as a matter of practicality, we cap the depth to which the tree can grow.

Figure 5 shows a portion of the contingency tree automatically generated for the Little Red Riding Hood domain. As before, inside the plan nodes solid arrows represent causal links and dashed arrows represent temporal constraints. The vertical i-beams alongside each plan node represent intervals during which exceptional actions can occur and result in inconsistencies that need to be handled. The arrows between plan nodes indicate which contingency narrative plan should be used if an inconsistency does in fact occur during interactive execution. The actual contingency tree for even the simple Little Red Riding Hood world requires thousands of contingencies, most of which are minor variations of each other (see Sect. 4.3 for execution details).

For online narrative plan execution, events in the current narrative plan are interpreted as abstract descriptions at the level commonly associated with plot. The events are used to generate directives to an underlying execution system that regulates game play. Each event can be thought of as a subset of the virtual world's overall state space. The execution system may or may not include semi-autonomous characters [34, 44]. We point the interested reader to details on the AUTOMATED STORY DIRECTOR framework [44] for specifics on one possible execution system.

4.3 Example: Little Red Riding Hood

Our approach to experience management, as implemented in the AUTOMATED STORY DIRECTOR framework [44], is illustrated in an interactive experience based loosely on the *Little Red Riding Hood* tale. The virtual world was built on a MOO (a text-based, object-oriented, multi-user dimension). Figure 6 shows a screenshot of the Little Red Riding Hood story in execution.

The player assumes the role of the Hunter (in the screenshot in Fig. 6, the player has chosen the name Fred). Although the Hunter is not the title character, the hunter is the character that ultimately "saves the day." Note that experience management

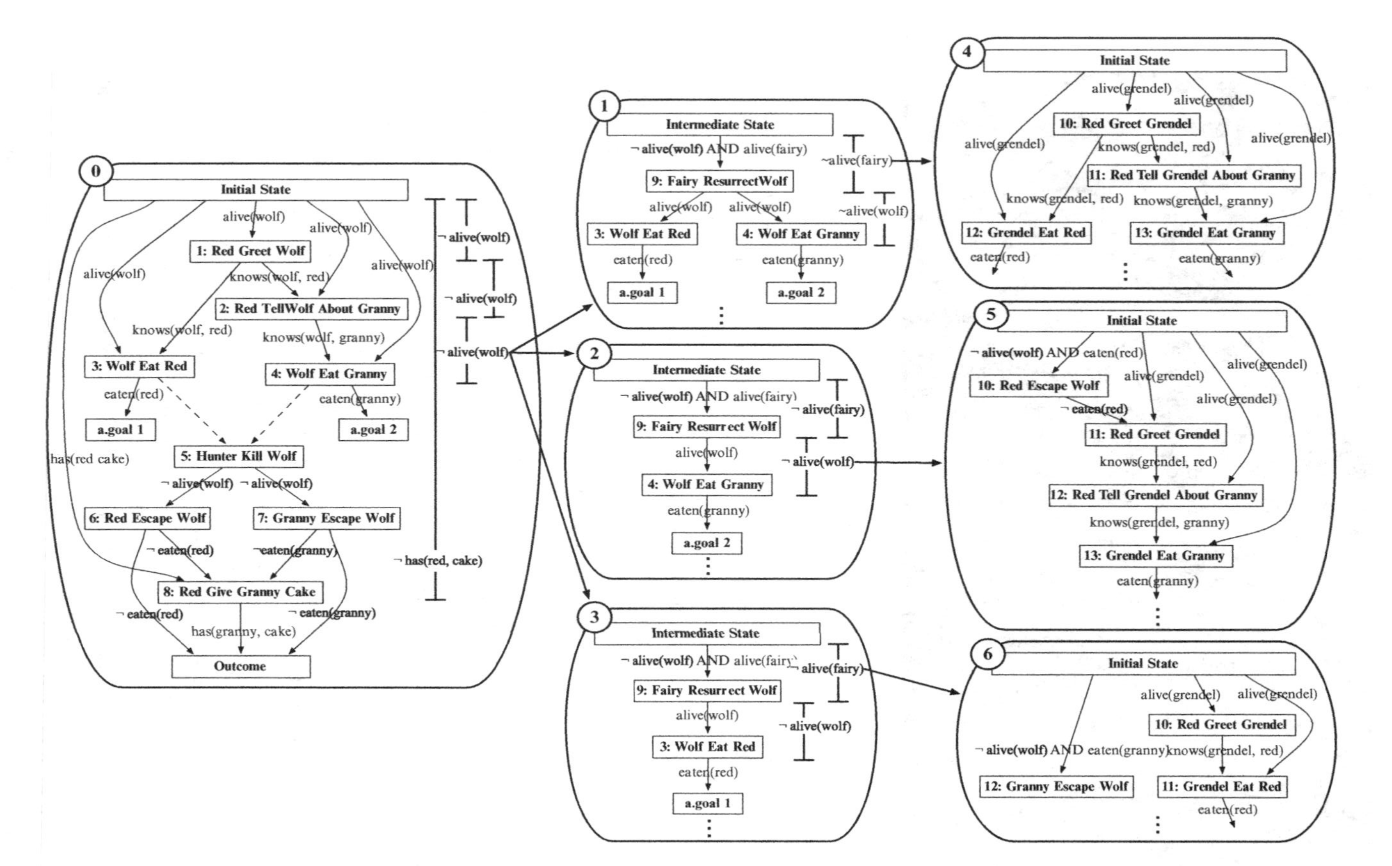

Fig. 5 A portion of the tree of narrative plan contingencies for Little Red Riding Hood

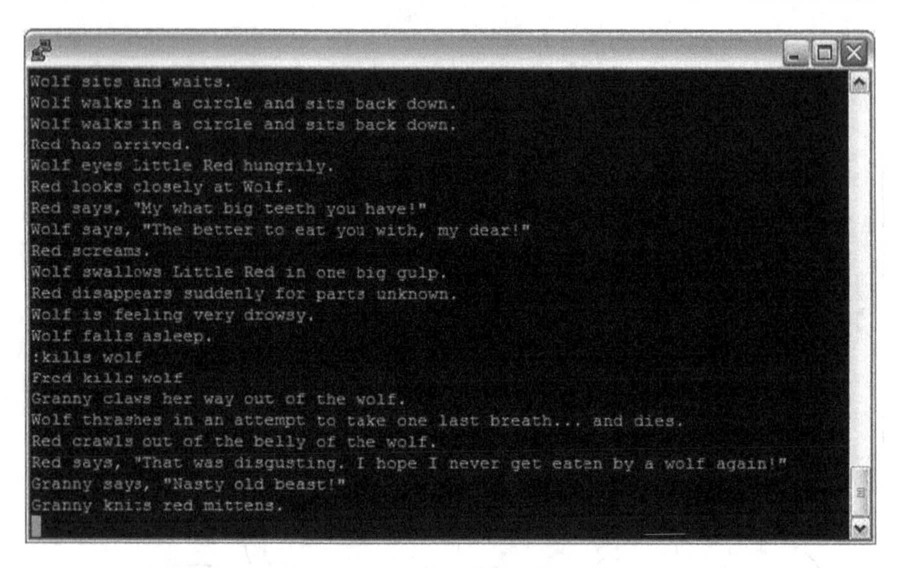

Fig. 6 Screenshot of the Little Red Riding Hood interactive story playing out in a MOO

can be performed regardless of which character the player controls. Experience management works best in rich virtual worlds with many characters. To make the Little Red Riding Hood domain more suitable for experience management, we extended the domain to include two extra characters: a fairy and a monster named Grendel. The fairy has the power to resurrect dead characters. Grendel, like the Wolf, is capable of swallowing other characters alive. In this simple example, the principal way in which the player expresses his or her agency is through the act of killing other characters at times other than that specified in the current narrative trajectory. A portion of the contingency narrative plan tree is shown in Fig. 5, only showing a few interesting branches (for space, plans are truncated to show only the actions that occur before the author goals). The exemplar narrative plan is the root of the contingency tree, shown at the left of the figure. Consider the narrative plan node labeled 1. To reach this trajectory, the player must create an inconsistency by killing the Wolf before it can eat either Red or Granny. The simplest alternative trajectory is to have the Fairy resurrect the Wolf, who then continues as normal. If for some reason the Fairy is also killed by the player, Grendel can fill the role of the character who eats Red and Granny, achieving the author goals. Note that in the exemplar narrative, the plot points specifying that the Wolf eat Red and that the Wolf eat Granny are unordered with respect to each other. This creates the possibility of multiple branches based on a race condition between the player's killing of the Wolf and the achievement of the two author goals: the Wolf can be killed before eating Red or Granny (contingency plan 1); the Wolf can be killed after eating Red but before eating Granny (contingency plan 2); or the Wolf can be killed after eating Granny but before eating Red (contingency plan 3). Each possible ordering of events in the race condition results in a slightly different narrative trajectory. The

Little Red Riding Hood domain and exemplar narrative shown in Fig. 2 result in 1,319 branches when the contingency tree is generated to a depth of 5 (the root of the tree, the exemplar narrative, is at depth 0). With a depth of 5, the contingency tree can handle five exceptional player actions in one play session before reverting to an emergent, unmanaged world. It takes approximately 43 min (approximately 11 min spent on garbage collection) to generate the contingency tree on an Intel Core2 Duo 3 GHz system with 3 GB of RAM and 100 GB of virtual memory running ALLEGRO CL® 8.0.

5 Player Modeling

Having considered the Experience Manager as a proxy for the designer, we may also consider the extent to which the player's preferences are part of the definition of "optimal" narrative trajectories.

Building and using models of player behavior is becoming increasingly prevalent in commercial video games, as doing so enables a computational game system to learn about the player to make decisions that impact the player's experience in a positive way. Player modeling in games has been used to maximize coherence, interest, and enjoyment.

- *Maximizing coherence of player experience.* A player model can be used to predict when the player might perform actions that diverge from the expected sequence and respond appropriately [23, 32].
- *Maximizing interest.* Learning player preferences over plot points and other narratively salient situations allows a game system to present an experience that is customized to the player's interests [4, 29, 50].
- *Maximizing enjoyment.* Learning player preferences over style of play has been shown to translate directly toward more engaging, enjoyable experiences [35,53].

The greater the Experience Manager's knowledge of its audience, the more informed its decisions about the player's experience will be. Due to the dynamic nature of games and virtual worlds, the point at which the Experience Manager has the *most* information about the player is just before a decision needs to be made. These facts motivate both learning and using a player model regularly during the course of the player's experience, and we present these tasks as two computational challenges for an Experience Manager to overcome: learning a profile of the player, and effectively using this model to positively affect the player's experience.

5.1 Learning About the Player

One promising approach, as implemented in the PASSAGE system [52, 53], is to learn about the player regularly throughout their interactive experience. One advantage of this approach, as opposed to learning about the player *before* game play begins, is that if the player's preferences change as the experience unfolds, the player model can be refined. Specifically, we propose to learn the player's preferences toward different styles of play [52], drawn from Laws' theory [27] for providing entertaining pen-and-paper role playing games. Laws identifies the following play styles:

- Fighter (f) – for players who enjoy engaging in combat
- Method Actor (m) – for players who enjoy having their personality tested
- Storyteller (s) – for players who enjoy considering complex plots
- Tactician (t) – for players who enjoy thinking creatively
- Power Gamer (p) – for players who enjoy gaining special items and abilities

Thus, a player model is a vector of scalars, $\langle f, m, s, t, p \rangle$, describing the extent to which the player has exhibited the traits of each play style. To determine whether the player is exhibiting a particular play style, player actions in the domain theory are annotated as being indicative of different styles of play; whenever the player performs an action that has been annotated, the corresponding value in the model increases. The player model is thus an estimate of the player's inclinations toward playing in each of the modeled styles.

5.2 *Using a Player Model*

Given the goal of maximizing player enjoyment, we can leverage the primary assumption of the PASSAGE system [52]: that players will enjoy events which allow them to play in their modeled play-styles more than events which favor other styles of play. Annotations on events (those performed by players and NPCs) indicating the play style that they are most suited for link the player model to the real-time execution of a narrative sequence. Thus the player model, represented as a vector of play style preference strengths, acts as a metric for each sequence to calculate its expected utility. This calculation could be as simple as examining the distribution of actions in the narrative sequence based on their annotations as to which play styles they support. For example, with a model of $\langle f = 1, m = 0, s = 0, t = 1, p = 2 \rangle$, the ideal narrative for this player would be made up of a collection of actions, distributed such that 25% appeal to fighters, 25% to tacticians, and 50% to power gamers. In the event that a narrative is not ideal for a player, the expected utility will be some value in the interval $[0, 1)$ indicating appropriateness based on event annotations.

Previously, we considered *how* to determine whether a player action is exceptional or not. We now consider *why* the player performs an exceptional action. There are many reasons why exceptions occur, including ignorance of the plotline, accident, malicious behavior (ie., trying to "break" the game),[4], or expression of a style

[4] The "cooperative contract" of interactive entertainment [61] suggests that if a player is not interested in being entertained in the way the game was designed, the designer need not be responsible for entertaining the malicious player.

of play that differs from the expected play style. If the exception occurs because of an expression of a particular play style, we wish to optimize the player's experience by repairing the narrative, accommodating the player's action into the narrative structure, and making any adjustments necessary to increase the expected utility of the subsequent narrative plan. Because the Experience Manager cannot know the precise configuration of the player model until the moment the exception is executed, narrative branches that account for many different possible configurations of the player model must be generated before the game being played.

Fortunately, the generation of narrative branches is one of the key features of the approach to experience management described earlier.

We extend our experience management approach to use the real-time dynamics of the player model in the following way. Instead of sequentially working through the four tiers of re-planning strategies (see Fig. 4a–d) escalating only when one of the strategies fails, the system executes all four strategy tiers for every possible inconsistency. Further, we modify tier (iv) to draw from many different sets of alternative author goals instead of selecting the next best set. We propose this because author goals force the story planner to consider trajectories that pass through different portions of state space [42]. In the absence of a human-level story generator, using sets of varied author goals forces the Experience Manager to explore a wider variety of trajectories, whereas without guidance, the planner may err on the side of making the fewest changes that it can.

The modifications described above result in one or more alternative branches for any given possible inconsistency. For example, if there are three sets of alternative author goals, then the maximum number of alternative branches per possible inconsistency is seven: one from tier (i), one from tier (ii), one from tier (iii), and three from tier (iv). See Fig. 7 for an illustration of branching execution incorporating a player model. The figure introduces decision nodes (diamonds) that select a child node (a narrative plan) based on the player model. In the illustration, killing the Wolf before event 1 has completed results in a possible inconsistency – the Wolf is unable to complete the action or any subsequent actions – and the possible inconsistency can be repaired in one of three ways.

When an inconsistency arises due to an exceptional player action, the system knows definitively that the current plan cannot continue to execute; the only non-determinism at this point is which branch to take. The system calculates the expected utility of each branch based on the configuration of the player model at the time the exceptional player action occurs. Thus the branch that actually begins execution at the time of the exceptional user action both restores causal coherence and tunes the player's experience according to his or her preferred play style.

The experience management approach incorporating real-time dynamics of the player model increases the number of narrative contingencies that must be generated *a priori*. However, the increase is by a linear number of branches per branching point and thus does not significantly increase the computational complexity of building the tree of contingency narratives. There is also an additional burden placed on the human designer in the sense that he or she must now provide as many sets of author goals as possible. Event templates for the world domain must also

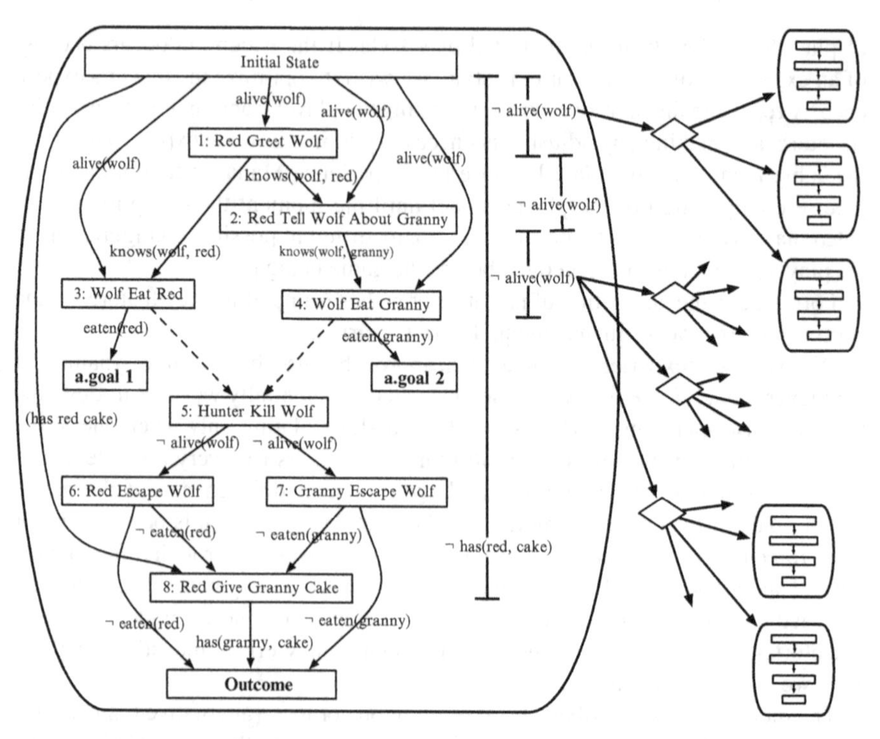

Fig. 7 A portion of a tree of narrative plan contingencies with decision nodes for inspecting the real-time player model

be annotated according to style of play. These additional authorial requirements are deemed relatively negligible, and future advancements in story generation will be less reliant on authorial guidance from humans. The added benefit is that there are multiple contingencies available for every possible inconsistency, meaning that the Experience Manager can optimize the narrative trajectory the player is on, with respect to the choices available.

The extensions to experience management, merging AUTOMATED STORY DIRECTOR and PASSAGE, have not been implemented; however, we believe the combination of story generation-based experience management and player model to be a promising way to address player agency and customization.

6 Conclusions

Experience management is the process whereby a player's agency is balanced against the desire to bring about a coherent, structured narrative experience. Intuitively, this is what game designers do when they construct a game world with a narrowly prescribed set of paths that deliver the player to a satisfying conclusion.

However, due to the growing trend toward greater player agency and greater content customization, we must consider computational approaches that offload design and management of the player's game play experience onto automated computational systems. In this work, we present an approach to automated, real-time experience management, in which we leverage the correlations between narrative and experience. By generating narrative trajectories that project possible experience into the future, a system is able to coerce a game or virtual world so that designer intent is preserved without diminishing player agency. The system is also able to reason about the narrative trajectory that maximizes the player's enjoyment based on acquired information about the preferences of the player toward certain styles of play. Thus, when the player exerts his or her agency in ways that are inconsistent with the provided narrative structure, the system is capable of seamlessly recovering *and* bringing the narrative trajectory in line with the player's inferred desires.

This perspective on how Artificial Intelligence can be used to create engaging gameplay expands the traditional role of an AI agent from adversarial opponent – focused on maximizing competitive payoff over time (e.g., beating the player) – to an agent with the goal of increasing the player's enjoyment. While there are many open research questions that remain with regard to generating better stories, the experience management framework suggests that whenever the global experience of a computer game is more important than achieving any one terminal state, be it a non-narrative game like chess or a game highly driven by plot, modeling the AI as storytelling is a beneficial approach.

References

1. Appling, D.S., Riedl, M.O.: The role of plot understanding in plot generation. In: S. Louchart, D. Roberts, M. Mehta (eds.) Intelligent Narrative Technologies II: Papers from the 2009 Spring Symposium (Technical Report SS-09-06). AAAI Press, Palo Alto, CA (2009)
2. Bae, B.C., Young, R.M.: A use of flashback and foreshadowing for surprise arousal in narrative using a plan-based approach. In: Proceedings of the 1st International Conference on Interactive Digital Storytelling (2008)
3. Bal, M.: Narratology: An Introduction to the Theory of Narrative. University of Toronto Press (1998)
4. Barber, H., Kudenko, D.: Dynamic generation of dilemma-based interactive narratives. In: Proceedings of the 3rd Artificial Intelligence and Interactive Digital Entertainment conference (2007)
5. Bates, J.: Virtual reality, art, and entertainment. Presence: The Journal of Tele-operators and Virtual Environments **1**(1) (1992)
6. Bates, J.: The role of emotion in believable agents. Communications of the ACM **37**(7), 122–125 (1994)
7. Blair, D., Meyer, T.: Tools for an interactive virtual cinema. In: R. Trappl, P. Petta (eds.) Creating Personalities for Synthetic Actors: Towards Autonomous Personality Agents. Springer (1997)
8. van den Broek, P.: The effects of causal relations and hierarchical position on the importance of story statements. Journal of Memory and Language **27**, 1–22 (1988)

9. Bruckman, A.: The combinatorics of storytelling: Mystery train interactive. Available: http://www.cc.gatech.edu/~asb/papers/combinatorics-bruckman-90.pdf (retrieved 1 Nov. 2007) (1990)
10. Bruner, J.: Acts of Meaning. Harvard University Press, Cambridge (1990)
11. Bruner, J.: The narrative construction of reality. Critical Inquiry **18**(1), 1–21 (1991)
12. Charles, F., Lozano, M., Mead, S., Bisquerra, A., Cavazza, M.: Planning formalisms and authoring in interactive storytelling. In: Proceedings of the 1st International Conference on Technologies for Interactive Digital Storytelling and Entertainment (2003)
13. Chatman, S.: Reading Narrative Fiction. Macmillan Publishing Company, New York (1993)
14. Cheong, Y.G., Young, R.M.: Narrative generation for suspense: Modeling and evaluation. In: Proceedings of the 1st International Conference on Interactive Digital Storytelling (2008)
15. Christian, D., Young, R.: Comparing cognitive and computational models of narrative structure. In: Proceedings of the 19th National Conference on Artificial Intelligence (2004)
16. Dennett, D.: The Intentional Stance. MIT Press, Cambridge, MA (1989)
17. Fitzgerald, A., Kahlon, G., Riedl, M.O.: A computational model of emotional response to stories. In: Proceedings of the 2nd Joint International Conference on Interactive Digital Storytelling (2009)
18. Gerrig, R.: Narrative thought? Personality and Social Psychology Bulletin **20**(6), 712–715 (1994)
19. Gerrig, R.J.: Experiencing Narrative Worlds: On the Psychological Activities of Reading. Yale University Press, New Haven (1993)
20. Graesser, A., Lang, K.L., Roberts, R.M.: Question answering in the context of stories. Journal of Experimental Psychology: General **120**(3), 254–277 (1991)
21. Graesser, A., Singer, M., Trabasso, T.: Constructing inferences during narrative text comprehension. Psychological Review **101**(3) (1994)
22. Grimm, J., Grimm, W.: Little red cap. In: Kinder- und Hausmärchen, 1st ed, vol. 1. Berlin (1812)
23. Harris, J., Young, R.M.: Proactive mediation in plan-based narrative environments. IEEE Transactions on Computational Intelligence and AI in Games **1**(3) (2009)
24. Jhala, A.H.: Cinematic discourse generation. Ph.D. thesis, North Carolina State University (2009)
25. Kelso, M., Weyhrauch, P., Bates, J.: Dramatic presence. Presence: The Journal of Teleoperators and Virtual Environments **2**(1) (1993)
26. Laurel, B.: Toward the design of a computer-based interactive fantasy system. Ph.D. thesis, Ohio State University (1986)
27. Laws, R.: Robin's laws of good GMing. Steve Jackson Games (2001)
28. Lebowitz, M.: Planning stories. In: Proceedings of the 9th Annual Conference of the Cognitive Science Society (1987)
29. Li, B., Riedl, M.O.: An offline planning approach to game plotline adaptation. In: Proceedings of the 6th Conference on Artificial Intelligence for Interactive Digital Entertainment Conference (2010)
30. Liden, L.: Artificial stupidity: The art of intentional mistakes. In: AI Game Programming Wisdom, vol. 2. Charles River Media (2003)
31. Loyall, A.B.: Believable agents: Building interactive personalities. Ph.D. thesis, School of Computer Science, Carnegie Mellon University (1997)
32. Magerko, B.: Player modeling in the interactive drama architecture. Ph.D. thesis, Computer Science and Engineering, University of Michigan (2006)
33. Mateas, M., Sengers, P.: Narrative intelligence. In: M. Mateas, P. Sengers (eds.) Narrative Intelligence: Papers from the 1999 Fall Symposium (Technical Report FS-99-01). AAAI Press, Menlo Park, CA (1999)
34. Mateas, M., Stern, A.: Towards integrating plot and character for interactive drama. In: K. Dautenhahn (ed.) Social Intelligent Agents: The Human in the Loop: Papers from the AAAI Fall Symposium (Technical Report FS-00-04). AAAI Press, Menlo Park, CA (2000)
35. Seif El-Nasr, M.: Interaction, narrative, and drama creating an adaptive interactive narrative using performance arts theories. Interaction Studies **8**(2) (2007)

36. McKoon, G., Ratcliff, R.: Inference during reading. Psychological Review **99** (1992)
37. Montfort, N.: Ordering events in interactive fiction narratives. In: B. Magerko, M. Riedl (eds.) Intelligent Narrative Technologies: Papers from the AAAI Fall Symposium (Technical Report FS-07-05). AAAI Press (2007)
38. Penberthy, J.S., Weld, D.S.: UCPOP: A sound, complete, partial-order planner for ADL. In: Proceedings of the 3rd International Conference on Knowledge Representation and Reasoning (1992)
39. Porteous, J., Cavazza, M.: Controlling narrative generation with planning trajectories: the role of constraints. In: Proceedings of the 2nd International Conference on Interactive Digital Storytelling (2009)
40. Prince, G.: A Dictionary of Narratology. University of Nebraska Press, Lincoln (1987)
41. Rattermann, M., Spector, L., Grafman, J., Levin, H., Harward, H.: Partial and total-order planning: evidence from normal and prefrontally damaged populations. Cognitive Science **25**, 941–975 (2001)
42. Riedl, M.O.: Incorporating authorial intent into generative narrative systems. In: S. Louchart, D. Roberts, M. Mehta (eds.) Intelligent Narrative Technologies II: Papers from the 2009 Spring Symposium (Technical Report SS-09-06). AAAI Press, Palo Alto, CA (2009)
43. Riedl, M.O., Saretto, C., Young, R.M.: Managing interaction between users and agents in a multi-agent storytelling environment. In: Proceedings of the 2nd International Conference on Autonomous Agents and Multi-Agent Systems (2003)
44. Riedl, M.O., Stern, A., Dini, D.M., Alderman, J.M.: Dynamic experience management in virtual worlds for entertainment, education, and training. International Transactions on System Science and Applications, Special Issue on Agent Based Systems for Human Learning and Entertainment **3**(1) (2008)
45. Riedl, M.O., Young, R.M.: An intent-driven planner for multi-agent story generation. In: Proceedings of the 3rd International Conference on Autonomous Agents and Multi-Agent Systems (2004)
46. Riedl, M.O., Young, R.M.: From linear story generation to branching story graphs. IEEE Journal of Computer Graphics and Animation **26**(3), 23–31 (2006)
47. Riedl, M.O., Young, R.M.: Narrative planning: Balancing plot and character. Journal of Artificial Intelligence Research **39** (2010)
48. Roberts, D.L., Riedl, M.O., Isbell, C.: Beyond adversarial: The case for game AI as storytelling. In: Proceedings of the 2009 Conference of the Digital Games Research Association (2009)
49. Rollings, A., Adams, E.: Andrew Rollings and Ernest Adams on Game Design. New Riders (2003)
50. Sharma, M., Mehta, M., Ontanón, S., Ram, A.: Player modeling evaluation for interactive fiction. In: Proceedings of the AIIDE 2007 Workshop on Optimizing Player Satisfaction (2007)
51. Swartjes, I., Theune, M.: A fabula model for emergent narrative. In: Proceedings of the 3rd International Conference on Technologies for Interactive Digital Storytelling and Entertainment (2006)
52. Thue, D., Bulitko, V., Spetch, M., Wasylishen, E.: Interactive storytelling: A player modelling approach. In: Proceedings of the 3rd Artificial Intelligence and Interactive Digital Entertainment Conference (2007)
53. Thue, D., Bulitko, V., Spetch, M., Webb, M.: Socially consistent characters in player-specific stories. In: Proceedings of the Sixth Artificial Intelligence and Interactive Digital Entertainment Conference (2010)
54. Trabasso, T., van den Broek, P.: Causal thinking and the representation of narrative events. Journal of Memory and Language **24**, 612–630 (1985)
55. Trabasso, T., Secco, T., van den Broek, P.: Causal cohesion and story coherence. In: H. Mandl, N. Stein, T. Trabasso (eds.) Learning and Comprehension in Text. Lawrence Erlbaum Associates (1984)
56. Trabasso, T., Sperry, L.: Causal relatedness and importance of story events. Journal of Memory and Language **24**, 595–611 (1985)
57. Weld, D.: An introduction to least commitment planning. AI Magazine **15** (1994)

58. West, M.: Intelligent mistakes: How to incorporate stupidity into your AI code. Game Developer Magazine – Digital Edition (2008)
59. Young, K., Saver, J.L.: The neurology of narrative. SubStance: A Review of Theory and Literary Criticism **30**, 72–84 (2001)
60. Young, R.M.: Notes on the use of plan structures in the creation of interactive plot. In: M. Mateas, P. Sengers (eds.) Narrative Intelligence: Papers from the AAAI Fall Symposium (Technical Report FS-99-01). AAAI Press, Menlo Park (1999)
61. Young, R.M.: The co-operative contract in interactive entertainment. In: K. Dautenhahm, A. Bond, L. Canamero, B. Edmonds (eds.) Socially Intelligent Agents. Kluwer Academic Press (2002)
62. Young, R.M., Riedl, M.O., Branly, M., Jhala, A., Martin, R., Saretto, C.: An architecture for integrating plan-based behavior generation with interactive game environments. Journal of Game Development **1**, 51–70 (2004)

Intelligent Machinima Generation for Visual Storytelling

Arnav Jhala and R. Michael Young

Abstract This chapter describes Darshak, an end-to-end system that automatically generates camera shot sequences for generation from a given story and visualization goals. The shot sequences that are generated by Darshak are visualized on a commercial 3D game engine automatically by an execution module implemented on the game engine.

1 Visual Storytelling: Motivation and Example

Through video is becoming increasingly prevalent in human culture. Last few decades have seen entertainment and educational applications involving interactive and non-interactive narratives, communicated through a variety of media including moving visuals in films, television, and more recently in video games. Experienced consumers of visual narration have learned established creative conventions for effective visual storytelling. A better understanding of the creation, consumption, and evaluation of cinematic styles is pertinent for creating tools that can assist in such a process, enhance human creativity in creating visual material, and use this medium for educational purposes. In this chapter, we present an intelligent movie directing and cinematography system – Darshak. Development and evaluation of Darshak provide insight into the process of identification and evaluation of visual cinematic devices (e.g., shot composition, camera transitions) and the creative process of cinematic reasoning through a computational algorithm that performs exploration of the creative space of these devices to automatically produce compelling visualizations. This enables directed systematic exploration of the large space of possible visual realizations of stories.

Unlike text, 2D art, and music, cinema is a relatively new art-form that has only received serious academic attention over the last half century. Storytelling through

A. Jhala (✉)
University of California Santa Cruz, Santa Cruz, CA 95064, USA
e-mail: jhala@cs.ucsc.edu

P.A. González-Calero and M.A. Gómez-Martín (eds.), *Artificial Intelligence for Computer Games*, DOI 10.1007/978-1-4419-8188-2_7,

the moving image has rapidly evolved from short silent films to CG composited SFX laden movies. Throughout this evolution, cinematic artists took advantage of the constraints of the technology of a given era and developed sophisticated idioms that had powerful impact on their viewers. More recently, cheap cameras, the internet distribution venues, and the ability to generate animation through low-cost game engines have generated a new breed of amateur video producers.

Film directors and cinematographers have identified and established creative conventions for effective visual storytelling. These conventions have been developed through an exploration of possibilities afforded by the constrained technologies (e.g., the physical film camera), and using computational methods (e.g., editing software, CG compositing) to create powerful experiences. Developers of each of these types of cinematic venues have experimented with principles of which some have been learned so well by the viewers that they define entire genres of stylistic filmmaking. For example, it is easy for a trained eye to identify a western themed film from the typical camera shots and transitions that are used during scenes that involve conflict among opposing parties. These conventions even hold when the setting is not of a typical western town, but is futuristic (e.g., Cowboy Bebop Anime). We will explore computational representations of such cinematic rules and algorithms that can reason about cinematic rules for creating compelling visualizations.

The use of patterns of shots and shot sequences in film has been well documented by cinematographers [1, 10]. These film idioms are understood by cinematographers and film audiences as stereotypical ways of communicating specific story elements or contexts. In recent work, we have begun the process of formalizing filmic communicative actions in terms of the actions' intentional constraints. This approach has been useful for selecting low-level camera operators for filming individual actions. This representation can be used to depict more abstract cinematic discourse structures intended to convey the rhetorical structure of a cinematic discourse (e.g., rising action, climax, denouement).

A number of researchers in computer graphics [2, 18] have addressed the issue of automating camera placement, though primarily from the point of view of geometric composition rather than cinematic discourse. The virtual cinematographer system developed by [5] models the shots in a film idiom as a finite state machine that selects state transitions based on the run-time state of the world. Tomlinson et al. [19] have used expressive characters for driving the cinematography module for selection of shots and lighting in virtual environment populated with autonomous agents. More recently, the use of neural network and genetic algorithm-based approaches in finding best geometric compositions [8, 9] has been investigated.

This work has been limited to the extent that the overall structure and flow of the cinematic discourse it produces is typically built from local geometric information about the setting. Decisions about shot selection are based, to a large extent, on the content of individual shots with little consideration to the relation each shot or shot sequence bears to its adjacent sequences. In situations where camera motion is made coherent, it is done through geometric means and not the context of the underlying narrative. Constructing cinematic discourse that attends to these inter-sequence intentional relationships is critical in narrative, where the dynamics of the situation model as the narrative unfolds is critical to the user's experience (as evidenced by the role of expectation and expectation violation noted above). This dependence on the careful construction of inter-sequence relationships is common to both the work we propose and much of the work on natural language discourse generation and is the motivation for our exploration of those techniques to produce cinematic discourse.

2 Darshak System Architecture

Chatman [3] proposed a bipartite representation of narrative. In Chatman's model, a narrative can be viewed as composed of two interrelated parts: the *story* and the *discourse*. The story describes the fictional world with all its content, characters, actions, events, and settings. The discourse contains medium-specific communicative content responsible for the *telling* of the narrative, for instance, a selection of a subset of events from the fabula, an ordering over these events for recounting, and linguistic communicative actions to tell the story. In this project, our focus is on representation and reasoning about elements at the discourse level, specifically narrative discourse that is communicated visually through the use of cinematic conventions by moving a virtual camera in a 3D environment rendered within a game engine.

Cinematic discourse in the Darshak system is generated by a hierarchical partial order causal link planner whose abstract actions represent cinematic patterns and whose primitive actions correspond to individual camera shots. As shown in Fig. 1, the system takes as input an operator library, a story plan, and a set of communicative goals. The operator library contains a collection of action operators

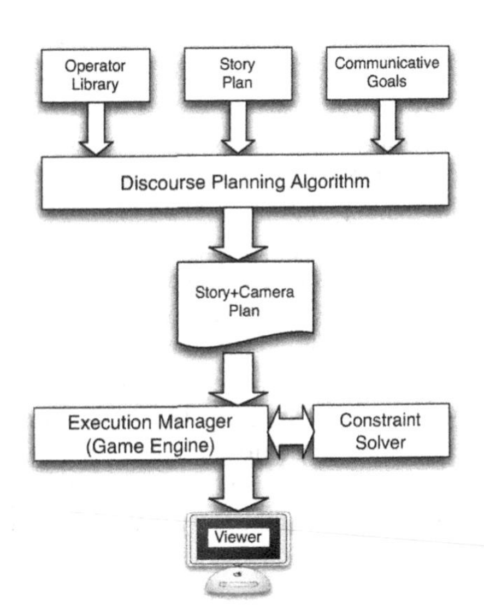

Fig. 1 Overview of Darshak's architecture

that represent camera placement actions, transitions, abstract cinematic idioms, and narrative patterns. Camera placement and transition actions, represented as primitive operators, affect the focus of visual they represent and have preconditions that encode continuity rules in cinematography. Operators representing abstract cinematic idioms and narrative patterns affect the beliefs of the viewers and encode recipes for sequencing primitive or abstract operators. A description of the story to be filmed is the input as a plan data structure that contains the description of the initial state of the story world, a set of goals achieved by the story characters' actions, and a totally ordered sequence of the story characters' actions and the causal relationships between them. The input story plan is added to the knowledge base for the discourse planner in a declarative form using first-order predicates that describe the elements of the data structure. The communicative goals given to the system describe a set of belief states to be established in the mind of the viewer.

The cinematic discourse planning algorithm performs both causal planning and temporal scheduling. To build a discourse plan, it selects camera operators from the operator library and adds them to the plan to satisfy specific communicative goals or preconditions of other communicative actions already in the plan. The algorithm binds variables in the camera operators, like start-time and end-time, relating the camera actions to corresponding actions in the story plan.

The output of the planning algorithm is a plan data structure containing a temporally ordered hierarchical structure of camera operators with all operator variables bound. The resulting plan is merged with the story plan and sent to a game engine for execution (Fig. 2).

2.1 An Example of Automated Machinima Generation in Darshak

The process of creation of a visualization plan in Darshak is illustrated by the following story about a thief, Lazarus Lane, who goes to a town in Lincoln County, Nevada to steal the tax money that is stored in the local bank. In the story, Lane successfully steals the money after winning Sheriff Bob's favor and being appointed as the deputy. The entire input story is shown in Table 1. The initial state for the discourse planning problem contains sentences about the story actions. The goal state contains the lone goal for the discourse planner [BEL V (HAS TAXMONEY LANE)]. It is assumed that the viewer has no prior beliefs about the story world. The discourse generated by the planner communicates to the viewer how Lane successfully steals the tax money from Lincoln county. As a discourse planner, Darshak operates on planning communicative actions that serve to manipulate the beliefs of viewers about the story. Darshak's communicative operators are designed to query conditions and actions in the story world that are related to the content of communication. In this example, the condition (HAS TAXMONEY LANE) will be queried for the actions in the story that lead to the establishment of that condition as well as the relevant causal chain of events leading up to this condition.

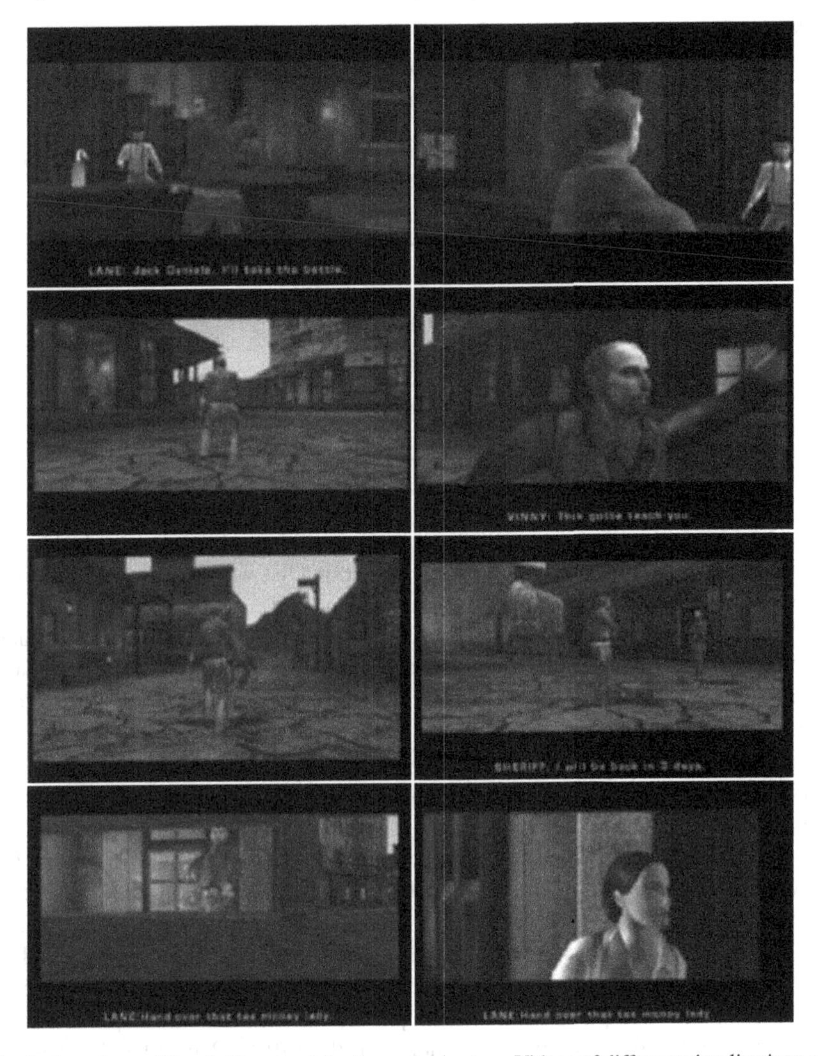

Fig. 2 Snapshots of Darshak's output for the example story. Videos of different visualizations can be found at: `http://www.youtube.com/user/cyclonurb`

Initially, the goal of the planning problem is established by instantiating a new discourse action in the plan: SHOW-ROBBERY (LANE, TAXMONEY, BANK). This action is chosen because it has an effect [BEL V (HAS LANE TAXMONEY)]. From the planner's perspective, the SHOW-ROBBERY action is causally motivated by the open condition of the goal state. The abstract action selected by the planner represents one of the narrative patterns that can be used to achieve the goal of telling the viewer the story of a character obtaining an object through a sequence of actions in

Table 1 Example story

Step 1 Lane goes to the Bar
Step 2 Lane asks the Bartender for a drink
Step 3 Lane overhears that Vinny the outlaw has murdered the town's deputy Sheriff
Step 4 Lane overhears that Sheriff Bob has cancelled the plans for going out of town as the town is without a deputy
Step 5 Lane goes to see Vinny
Step 6 Vinny threatens Lane
Step 7 Lane Shoots Vinny
Step 8 Lane goes to Sheriff
Step 9 Sheriff Bob appoints Lane as Deputy Sheriff
Step 10 Sheriff Bob leaves town
Step 11 Lane goes to the bank
Step 12 Lane Threatens the Teller
Step 13 Teller gives Lane the tax money from the bank vault

the story-world.[1] Given this choice, the planner instantiates the action and adds it to the discourse plan, and updates the open conditions list with the pre-conditions of the SHOW-ROBBERY action.

The abstract action is then decomposed in the next planning step into constituent actions. In this example, the SHOW-ROBBERY action is expanded to three sub-actions: SHOW-ROBBERY-ACTION, SHOW-THREAT, and SHOW-RESOLUTION. The discourse actions are bound to story actions through the instantiation of constraints on the operators. The SHOW-ROBBERY-ACTION, in this case, has constraints that bind the story step that this action films. The step in the story plan that indicates successful robbery is Step 13, where the teller gives Lane the tax money from the vault. This action has the effect (HAS LANE TAXMONEY). The corresponding constraint on the SHOW-ROBBERY-ACTION operator is [EFFECT ?S (HAS ?CHAR ?OBJ)] which binds ?S to STEP13, ?CHAR to LANE, and ?OBJ to TAXMONEY. In this way, constraints on the discourse operators are checked by queries into the knowledge base describing story plan; as a result, correct bindings for the story world steps are added to the plan structure. Once the step is correctly bound, the start and end temporal variables for the SHOW-ROBBERY-ACTION are bound to (START STEP13) and (END STEP13), respectively, and the temporal constraint graph is updated with these new variables. The planning process continues with expansion of abstract actions and addition of new actions to satisfy open conditions.[2] Figure 3 illustrates a complete camera plan.

During the addition of temporal constraints to the discourse plan, the planner maintains a simple temporal graph [20] of all the steps' time variables and checks

[1] At this point, the planner could have chosen any of the other patterns with the effect [BEL V (HAS ?CHARACTER ?OBJECT)] that would have also satisfied the story world's constraints.

[2] As part of the planning process, the planner creates an entire *space* of possible plans that might solve its planning problem. The example here traces a single path through the construction process leading to the correct solution.

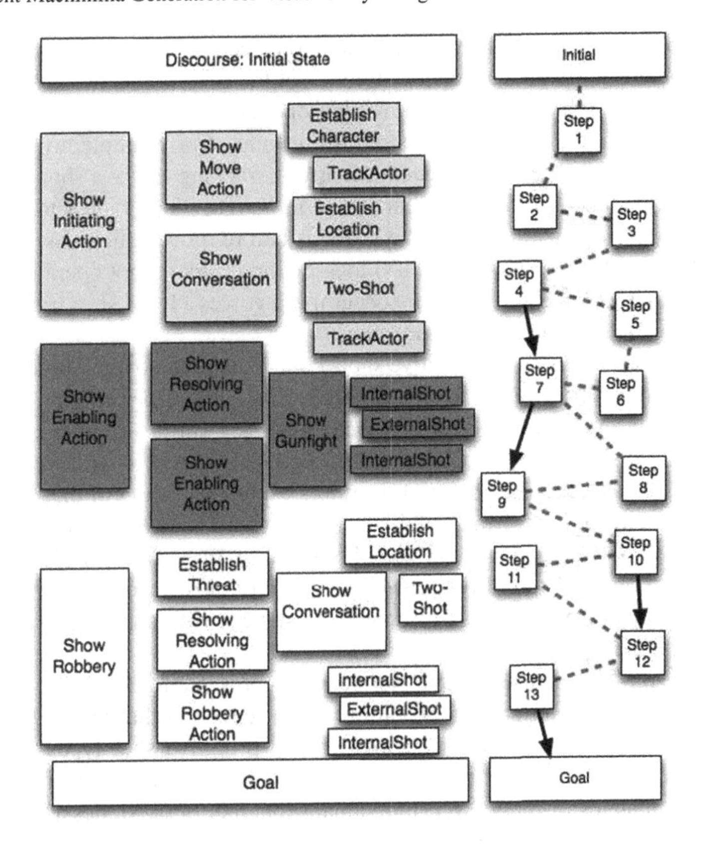

Fig. 3 Fully realized discourse plan for the example story. The story plan steps are shown on the *right*. The discourse plan actions are shown hierarchically from *left* to *right*. *Right most* actions are primitive actions and actions to the *left* are abstract actions. Primitive camera actions film the story world actions that they are adjacent to in the figure. *Dotted red lines* show the temporal sequence of actions from top to bottom and *black arrows* depict causal links between story world actions

for consistency of the graph after the addition of each action. Each temporal variable has a protection interval during which it satisfies all the constraints imposed on it by the camera operators. As actions are added to the plan, the temporal consistency checking algorithm constantly updates this protection interval and checks for invalid intervals to prune inconsistent temporal orderings for the actions.

2.1.1 Representation of Stylistic Elements

Cinematographers typically make use of a relatively small number of primitive camera shots [1]. From a cinematographer's perspective, primitive shot specifications describe the composition of the scene with respect to the underlying geometric context. The main goal of a cinematographer when choosing a primitive shot is to compose the shot such that the viewer focuses on a certain aspect of the scene that is being framed. Through a sequence of coherent focus shifts, scenes are built that communicate the story. Represented as plan operators, primitive shot definitions capture certain coherence and shot-composition rules. For example, to avoid disorienting the viewer, one of the preconditions of a tracking shot – a shot where the camera moves relative to the movement of the view target – is that the actor or object that the operator is tracking should be in focus before movement starts. Adding a precondition (INFOCUS ?F)@[TSTART) to a tracking camera shot operator ensures that a jump to tracking camera will not occur if it involves a focus shift from another object or from a different shot composition.

Composition and focus of primitive shots contribute to the overall dramatic presentation of the story. Van Sijll, in her book *Cinematic Storytelling* [17], lists 100 camera shot conventions that are used to convey various dramatic elements in movies. Each shot's dramatic value is described with respect to its screenplay and blocking. Darshak's representation of operators captures a subset of these dramatic effects of camera actions. Nine primitive operators and several variations of these operators are implemented in the system. For instance, three variants of the LookAt operator are LookAt-Close, LookAt-Medium, and LookAt-LongShot, which view an actor or object from progressively farther distances with appropriate cinematic framing for each distance.

Film idioms lend themselves to a hierarchical representation, with sub-parts that are themselves characterizations of idioms and a reduction that terminates in the types of primitive shots described above. While primitive shot types determine viewer's focus, abstract shot types represent the effect of focus and focus shifts on the mental states of the viewer. Individual shots like LookAt, described above, have a denotative meaning associated with them that focus the viewer's attention on elements in the frame. Abstract shot types also encode relationships *between* primitive operators and provide a mechanism for expressing established idioms as specific recipes or decompositions. Abstract plan operators, such as those used in typical HTN-style planning [16], are used in Darshak to capture such communicative phenomena by explicitly representing the different ways in which sequences of more primitive shots can convey discourse-level information about the story being filmed.

There are established patterns of effective storytelling that have been documented by narrative theorists [6, 14]. These patterns implicitly manipulate focus of attention in effective ways. These narrative patterns are also operationalized as plan operators in Darshak. This representation allows the operator designer to specify constraints on salient elements according to their preferences. It is possible to describe narrative patterns as hierarchical plan operators. The parameters of the pattern are captured using the parameters of the operator. The role of each parameter within the context of the story is represented by constraints on the parameters of the operator. The variations of each pattern can be expressed in terms of collections of alternative decomposition operators, all decomposing the same abstract action but each with varying sets of applicability constraints.

3 DPOCL-T Algorithm for Generating Cinematic Discourse

In addition to their use in generating plans for physical activity (e.g., robot task planning), planning algorithms have been successfully used in the generation of effective textual discourse [12] as well as for story generation [15]. As described in the previous section, the representation of narrative discourse operators in Darshak encodes a rich formal representation of the causal structure of the plan. Each dependency between goals, preconditions, and effects is carefully delineated during the plan construction process. The system searches through the space of all possible plans during the construction process and thus can characterize the plan it produces relative to the broader context of other potential solutions to its planning problems.

To build Darshak, we extended our earlier work [21] on the DPOCL hierarchical planning algorithm to create Decompositional Partial-Order Causal Link planning algorithm with Temporal constraints (DPOCL-T). DPOCL-T forms the core planning algorithm used in Darshak.[3] The following section provides formal definitions of the constituent parts that make up a DPOCL-T planning problem.

3.1 Action Representation

DPOCL-T supports durative actions with temporal constraints on temporal variables. Actions are defined using a set of action schemata consisting of an action type specification, a set of free object variables, a set of temporal variables, a set of temporally indexed preconditions of the action, a set of temporally indexed effects, a set of binding constraints on the variables of the action, and a set of temporal constraints on the temporal variables of the action.

This action representation differs from previous approaches to discourse planning in its explicit representation of temporal variables and constraints on these variables. The set of temporal variables implicitly contains two distinguished symbols that denote the start and end of the action instantiated from the schema in which they occur.

Actions that are directly executable by a camera are called *primitive*. Actions that represent abstractions of more primitive actions are called *abstract* actions. Abstract actions have zero or more *decomposition schemata*; each decomposition scheme for a given abstract action describes a distinct recipe or sub-plan for achieving the abstract action's communicative goals.

---

[3] Space limitations prevent us from providing a full discussion of the DPOCL planning algorithm. For more details, see [21].

3.2 *Domain, Problem, and Plan*

A planning *problem* in DPOCL-T specifies a starting world state, a partial description of desired goal state, and a set of operators that are available for execution in the world. Since conditions are temporally indexed, the initial state of the *camera* problem not only specifies the state of the world at the beginning of execution but also indicates the *story* actions and events that occur in the future (i.e. during the temporal extent of the camera plan). As a result, DPOCL-T is able to exploit this knowledge to generate camera plans that carefully coordinate their actions with those of the unfolding story. DPOCL-T plans are similar to those built by DPOCL except for the manner in which temporal ordering is specified. In DPOCL, the relative ordering of steps is expressed through explicit pair-wise ordering links defining a partial order on step execution. In DPOCL-T, the ordering is implicitly expressed by the temporally constrained variables of the steps.

3.3 *DPOCL-T Algorithm*

Given a problem definition as described in the previous section, the planning algorithm generates a space of possible plans whose execution starting in the problem's initial state would satisfy the goals specified in the problem's goal state. The DPOCL-T algorithm generates plans using a refinement search through this space. The algorithm is provided in Fig. 4. At the top level, DPOCL-T creates a graph of partial plans. At each iteration, the system picks a node from the fringe of the graph and generates a set of child nodes from it based on the plan it represents. Plans at these child nodes represent single-step refinements of the plan at their parent node. Plan refinement involves either causal planning, where steps' preconditions are established by the addition of new preceding steps in the plan, or episodic decomposition, where sub-plans for abstract actions are added to the plan. The final step of each iteration involves checks to resolve any causal or temporal inconsistencies that have been added to the child plans during plan refinement. Iteration halts when a child plan is created that has no flaws (i.e., that has no conflicts, no preconditions that are not causally established, and no abstract steps without specified sub-plans).

3.3.1 Causal Planning

In conventional planning algorithms, an action is added to a plan being constructed just when one of the action's effects establishes an unestablished precondition of another step in the plan. To mark this causal relationship, a data structure called a causal link is added to the plan, linking the two steps and that condition that holds between their relative execution times.

DPOCL-T ($P_C = \langle S, B, \tau, L_C, L_D \rangle, \Lambda, \Delta$)

Here P_C is a partial plan. Initially the procedure is called with S containing placeholder steps representing the initial state and goal state and τ containing temporal constraints on the initial and goal state time variables.

Termination: If P_C is inconsistent, fail. Otherwise if P_C is complete and has no flaws then return P_C

Plan Refinement: Non-deterministically do one of the following

1. Causal planning
   a. Goal Selection: Pick some open condition p@[t) from the set of communicative goals
   b. Operator Selection: Let S' be the step with an effect e that unifies with p. If an existing step S' asserts e then update the temporal constraints for t to be included in the protection interval for the effect e. If no existing step asserts e then add a new step S' to the plan and update the temporal constraint list with the variables introduced by step S'
2. Episode Decomposition
   a. Action Selection: Non-deterministically select an unexpanded abstract action from P_C
   b. Decomposition Selection: Select a decomposition for the chosen abstract action and add to P_C the steps and temporal and object constraints specified by the operator as the subplan for the chosen action.

Conflict Resolution:
For each conflict in P_C created by the causal or decompositional planning above, resolve the conflict by nondeterministically chosing one of the following procedures:

1. Promotion: Move S1 before S2 (Add constraints on the start and end time points of S1 and S2)
2. Demotion: Move S2 before S1 (Add constraints on the start and end time points of S1 and S2)
3. Variable Separation: Add variable binding constraints to prevent the relevant conditions from unifying

Recursive invocation: Call **DPOCL-T** with the new value of P_C.

Fig. 4 Sketch of the camera planning algorithm

In DPOCL-T, causal planning also requires enforcing temporal constraints on the intervals in which the relevant preconditions and effects hold. When Darshak establishes a causal link between two steps s_1 and s_2, additional constraints are also added to the plan. Suppose that s_1 (ending it execution at time t_1) establishes condition p needed by a precondition of s_2 at time t_2. When the causal link between s_1 and s_2 is added, Darshak also updates its constraint list to include the constraint $t_1 \leq t < t_2$. Further, because p must now be guaranteed to hold between t_1 and t_2, Darshak checks at each subsequent iteration that this constraint holds. In this case, we adopt terminology used by [13] and call the interval $[t_1, t)$ a *protection interval* on the condition p_i.

3.3.2 Decompositional Planning

Decomposition schemata in DPOCL-T are similar to their counterparts in DPOCL, both in their content and in their manner of use. The sole point of difference is in the schemata's representation of temporal constraints – rather than include a set of explicit pair-wise orderings between steps within a decomposition schema, DPOCL-T specifies partial orderings via constraints on the time variables determining the decomposition's constituent steps' execution.

3.3.3 Management of Temporal Constraints

As new steps are added to a plan being built, the effects of these steps may threaten the execution of other steps in the plan. That is, their effects might undo conditions established earlier in the plan and needed by the preconditions of some steps later in the plan. In DPOCL-T, a global list of temporal constraints is maintained as a directed graph of all the time variables involved in the plan's current steps. Whenever the planner adds a step, all the step's time variables are added to the directed graph, and the graph is completed with the step's new constraints representing the edges of the graph. Another step's precondition is threatened by a newly added step just when the effect of the new step changes the needed condition within its protection interval. Such threats are detected by identifying a cycle in the temporal constraints graph.

The Darshak algorithm produces plans offline from given stories, and we are concerned with the quality of generated plans rather than fast execution of the algorithm. The exponential explosion of the plan space is mitigated by careful design of operators with variable binding and temporal constraints. For complex domains, the hierarchical representation of operators makes this approach scalable. There is a significant authorial burden in writing domain specifications that can be eased through storyboarding interfaces.

3.4 Camera Control on the Game Engine

A virtual cinematography module, represented by several classes containing cinematic rules and camera placement commands, is responsible for managing execution of story and camera plans on the game engine. The high-level camera directives that are generated by the camera planner are translated to geometric constraints for input to a constraint solver for camera placement. The translator accepts the abstract annotated action sequence and translates its contents into a set of geometric constraints characterizing the best shot-composition given (a) the geometry of the world and (b) the position of the shot's characters. The translator uses the cinematic rules to generate constraints on the composition of the camera's frame. These constraints are input to a constraint solver that computes a heuristically best position of the camera, given the geometry of the world and the constraints that are required for a cinematic presentation of the shot (Fig. 5).

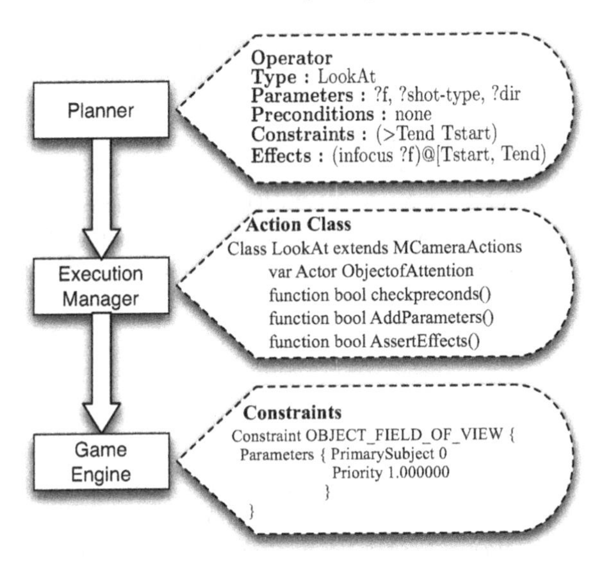

Fig. 5 Representation of camera actions in different modules

The rules of composition on the game engine are described in the form of geometric constraints. The forms and values of constraints are motivated by the attributes of the setting as specified by the designer, and these are related to the actual properties of a camera such as location, field-of-view, and orientation.

3.4.1 Constraint Solver

Given a director's shot request is expressed in the form of visual composition constraints, the system positions the camera to satisfy the given constraints in the context of the given 3D virtual scene. The desired visual message or camera shot is expressed in the form of constraints on how subjects appear in the frame. The constraint solver then attempts to find values for each camera parameter so that all given constraints are satisfied within a heuristically optimal degree to their respective optimal or preferred values. The camera constraint system in Darshak supports 15 different types of constraints on camera attributes in relation to the scene and participants. Constraints include framing a subject in the camera's field of view, viewing a subject from a desired vantage angle, excluding or obscuring an undesired subject from the frame, avoiding occlusions, and a variety of constraints on how the subject's image is projected into the frame. Any constraint that can be applied to an object can also be applied to a designated part of the object's geometry. The scene and object geometry is represented using a combination of oriented bounding boxes and binary space partition trees. These geometric data structures are used to estimate projections of subjects in the frame, fraction of occlusion between subjects, and to keep the camera inside the walls and architectural features of a building interior.

An execution monitor, which is implemented on the Unreal Tournament (UT) game engine, manages the execution of actions communicated by the server. It receives both the camera actions and story world actions. The primitive camera action classes (close-up, medium-shot, long-shot, track-actor, pan-actor-to-actor, internal-shot) use the cinematographer object to set up constraints (location, orientation, movement, and lens) for the player's camera. The player's view is updated after getting recommendation for the camera position from the cinematographer, which takes into consideration the currently set up constraint values for updating the camera location.

4 Empirical Evaluation

To evaluate the effects of different visualization strategies, three visualizations of the same story were prepared: one with a fixed camera position within the setting, one with an over-the-shoulder camera following the protagonist, and one with camera plan automatically generated by Darshak. The purpose for running these experiments was twofold. First, to investigate whether visualization strategies do indeed affect comprehension. Second, to evaluate the quality of visualization generated by Darshak using a representation of camera shots as communicative actions. The objective of the experiments was to determine whether visualizations generated by Darshak are coherent (as measured by viewers' perceptions of the attributes of the underlying stories).

Our experimental approach was to first map the stories being shown to subjects into a representation previously developed and validated as a cognitive model of narrative. We then probed subjects' understanding of the narrative that was presented in a cinematic to determine how closely their understanding aligned with the cognitive model's predictions about a viewer's mental model. As the underlying story elements in this system were defined as plan data structures themselves, the experimental design was based on previous work [4] relating these data structures to the mental models that users form during comprehension. To do this, a mapping is defined from plan data structures onto a subset of the conceptual graph structures adopted from the QUEST model, developed by Graesser et al. [7]. In the QUEST model [7], stories are represented as conceptual graph structures containing concept nodes and connective arcs. Together with a specific arc search procedure, QUEST was originally used to provide a cognitive model of question–answering in the context of stories (supported question types include why, how, when, enablement, and consequence questions). In our work, we make use only of the graph structures, referred to here as QUEST Knowledge Structures (or QKSs). They describe the reader's conception of narrative events and their relationships. The composition of nodes and arcs in a QKS structure reflects the purpose of the elements they signify in the narrative. For instance, if nodes A and B are two events in a story such that A causes or enables B, then A and B are represented by nodes in the QKS graph and are connected by a consequence type of arc.

Techniques used by Graesser et al. to validate the QUEST model were based on goodness-of-answer (GOA) ratings for question–answer pairs about a story shown to readers. Subjects were provided with a question about the story and an answer to the question, then asked to rate how appropriate they thought the provided answer was given the story that they had read. GOA ratings obtained from their subjects were compared to ratings predicted by the arc search procedures from QUEST model. Results from their experiments clearly showed a correlation between the model and the cognitive model developed by readers to characterize stories.

The algorithm for converting a Partial-Order Causal Link (POCL) plan data structure to the corresponding QKS structure is adopted from Christian and Young [4]. In this experiment, first the story, represented as a plan data structure, is converted to a corresponding QKS structure. Predictor variables proposed in the QUEST model are used to calculate predictions for the GOA ratings – the measure of goodness of answer for a question/answer pair related to the events in the story. These GOA ratings are compared against data collected from participants who watch a video of the story filmed using different visualization strategies. The three predictors that are correlated to the GOA ratings are arc search, constraint satisfaction, and structural distance. Correspondence between GOA ratings indicated by these predictor variables and by human subjects would be taken as an indication that the cinematic used to tell the story had been effective at communicating its underlying narrative structure.

4.1 Method

4.1.1 Design

To create story visualizations, we used two stories (S1 – shown in Table 1, and S2 – not shown) and three visualization strategies for each story (V1 – fixed camera, V2 – over-the-shoulder camera angle, and V3 – Darshak driven camera), creating six treatments. Treatments were identified by labels with story label as prefix followed by the label of the visualization. For instance, S2V1 treatment would refer to a visualization of the second story (S2) with fixed camera angle strategy (V1). Participants were randomly assigned to one of six groups (G1–G6). Thirty participants, primarily undergraduate and graduate students from the Computer Science Department at NC State University, participated in the experiment. Each participant was first shown a video and then asked to rate question–answer pairs of three forms: how, why, and what enabled. The process was repeated for each subject with a second video.

A Youden squares design [11] was used to distribute subject groups among our six treatments. This design was chosen to account for the inherent coherence in the fabula and to account for the effects of watching several videos in order. Assuming a continuous response variable, the experimental design, known as a Youden square, combines Latin Squares with balanced, incomplete block designs (BIBD). The Latin Square design is used to block on two sources of variation in complete blocks. Youden squares are used to block on two sources of variation – in this case, story and group – but cannot set up the complete blocks for latin squares designs. Each row (story) is a complete block for the visualizations, and the columns (groups) form a BIBD. Since both group and visualization appear only once for each story, tests involving the effects of visualization are orthogonal for those testing the effects of the story type. The Youden square design isolates the effect of the visual perspective from the story effect.

Each story used in the experiment had 70 QKS nodes. Of the 70 QKS nodes, ten questions were generated from randomly selected QKS elements and converted to one of the three question types supported by QUEST: how, why, and what enabled. For each of the ten questions, approximately 15 answer nodes were selected from nodes that were within a structural distance of three in the QKS graph generated from the story data structure. These numbers were chosen to have similar magnitude to Christian and Young's previous experiments, for better comparison.

4.1.2 Procedure

Each participant went through three stages during the experiment. The entire experiment was carried out in a single session for each participant. Total time for a single participant was between 30 and 45 min. Initially, each participant was briefed on the experimental procedure and was asked to sign the consent form. They were then asked to read the instructions for participating in the study. After briefing, they watched a video of one story with a particular visualization according to the group assignment (Table 2). For each video, users provided GOA ratings for the question–answer pairs related to the story in the video. Participants were asked to rate the pairs along a four-point scale: good, somewhat good, somewhat bad, and bad. This procedure is consistent with earlier experiments [4, 7]. Next, they watched a second video with a different story and visualization followed by a questionnaire about the second story. The videos were shown in different orders to common groups to account for discrepancies arising from the order in which participants were shown the two videos.

4.2 Results and Discussion

The mean overall GOA ratings recorded for the two stories are shown in Table 3 along with the standard deviations. These distributions of GOA scores do not present

Table 2 2×3 Youden squares design for the experiment. G1 through G6 represent six groups of participants with five members in each group. They are arranged so that each story and visualization pair has a common group for other visualizations

Viz	Master shot	Over the shoulder	Darshak
S1	G1,G4	G2,G5	G3,G6
S2	G5,G3	G6,G1	G4,G2

Table 3 Mean GOA ratings and standard deviations from the experiment

GOA (SD)	V1	V2	V3
S1	1.69 (±0.91)	1.74 (±0.82)	1.70 (±0.79)
S2	1.76 (±0.59)	1.51 (±0.67)	1.78 (±0.59)

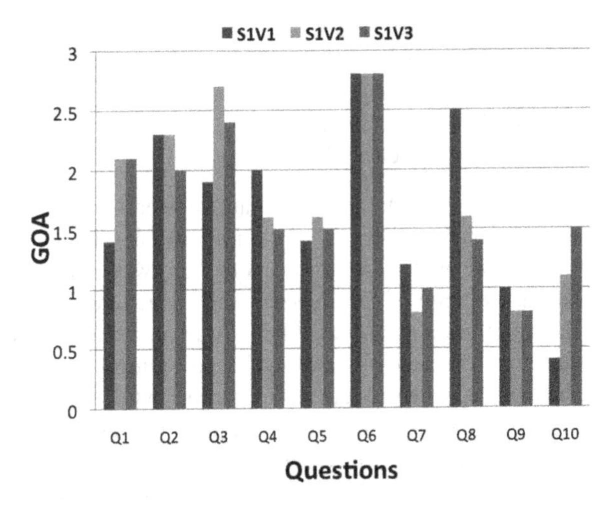

Fig. 6 GOA ratings for Story 1 across the three visualization strategies (S1 and S2 are stories, V1 – Master Shot, V2 – Over-the-Shoulder shot, and V3 – Darshak are visualization strategies)

any problem for multiple regression analyses as the means do not show ceiling or floor effects. The standard deviations are high enough to rule out the potential problem of there being a restricted range of ratings. GOA ratings were recorded along a four-point scale (Strongly Disagree, Disagree, Agree, Strongly Agree) for each question–answer pair. This was done to match the scale used in the original QUEST experiments. The GOA numbers shown in Table 3 indicate on preliminary observation that the GOA ratings for V1(Master Shot) and V3 (Darhsak) are significantly correlated with V2 (Over-the-Shoulder shots). The standard deviations for V3 are lower than the other treatments in both stories. This indicates that participants converge better on rating questions in Darshak-generated visualizations.

An interesting observation for V2 is that in story 2 the mean GOA ratings are significantly lower than the other two treatments with a significantly high standard deviation. These numbers support the intuition that participants form their own interpretation of events in the story while looking at shots that are over-the-shoulder leading to the wide disparity in ratings in going from story 1 to story 2. While mean ratings provide an overall idea of the participant's responses, GOA ratings for individual questions across different visualizations provide more insights into the differences across visualizations. Figure 6 summarizes mean GOA ratings for individual questions related to story 1 for the three visualization treatments. Question numbers 1, 8, and 10 are particularly interesting as there is a big variation in the

GOA ratings for the master shot visualization and the other two treatments, which have quite similar ratings. The question–answer pairs in discussion here are presented below:

1. Question: Why did Lane challenge Vinny?
   Answer: Because he wanted to kill Vinny.
8. Question: Why did Lane challenge Vinny?
   Answer: Because Lane wanted to steal tax money.
10. Question: Why did Lane meet Sheriff Bob?
    Answer: Because Lane needed a job.

In Q1 and Q10, the ratings for V1 are significantly lower. This could be explained by examining the relationships between the question–answer nodes. In all three cases, the question answer nodes are two or more arcs away in distance along the causal chain of events. In case of the arc-search and structural distance predictors from QUEST, these are good answers as they do lie on a causal chain of events leading to the question. The necessity and sufficiency constraints in the constraint satisfaction predictor reduce the strength of the answer. In Q1, for example, it is not necessary for Lane to challenge Vinny. He could just shoot him right away. In this case, viewers who were familiar with the gunfight setting chose to label the challenge as being an important step in killing Vinny as, for them, it forms an integral part of the gunfight sequence. In the master-shot visualization, the gunfight sequence was not even recognized as a gunfight by most participants (information obtained from post-experiment interview). This analysis indicates that additional consideration of the "why" type of questions on other nodes is needed to determine the effects of visualization strategies on GOA ratings related to perceived causal connections between events in the story.

Figure 7 shows the average ratings for each question for the second story. The interesting responses are the ones that have a significant variation in mean

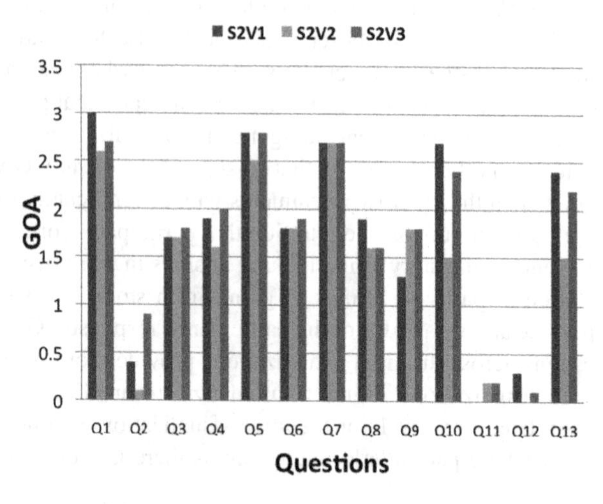

Fig. 7 GOA ratings for Story 2 across the three visualization strategies (S1 and S2 are stories, V1 – Master Shot, V2 – Over-the-Shoulder shot, and V3 – Darshak are visualization strategies)

ratings across different visualizations. In this story, unlike the data for story 1, the differences between ratings were relatively smaller. The interesting observations, however, were the ones where one of the treatments rated the answer as a "bad" answer (rating <1.5) and the other treatments rated the answer as a "good" answer (rating >1.5).

Post-experiment interviews were carried out after participants completed the rating forms for both stories. During these interviews, the participants were asked to give subjective answers to questions about quality of the story and videos. The main purpose of these questions was to get additional information about metrics that were not considered in previous approaches but may play a role in the analysis of GOA ratings. The data collected from this survey provide insight into a possible extension of the cognitive model of story understanding that takes into account features of discourse that are not currently represented. Based on the subjective data, a majority of the subjects preferred system-generated videos to the other two videos. Most users reported that visualization did affect their engagement in the story. System-generated visualizations were rated as being more engaging. Users preferred camera movements over static camera shots as they perceived the scene to be more dynamic and interesting. While these qualitative responses are hard to measure statistically, they do point to uniformity among experiment participants regarding their preference for system-generated visualizations over other visualization strategies.

There are several caveats to the evaluation strategy based on cognitive models of story comprehension. The results of the study only address the cognitive aspect of story perception and do not directly measure specific discourse effects and aesthetic quality of produced output. Further work is needed within a richer story domain with opportunity to exploit subtleties of visual communication to fully evaluate the potential of Darshak's ability of generating aesthetic sequences. The current focus for Darshak, however, is on generating coherent sequences. A sound evaluation of coherence presented in this chapter establishes Darshak's success for this focus.

5 Conclusion

In this chapter, we have presented the design, development, and evaluation of a system that plans cinematic visualizations of stories. As games move toward dynamic environments, props, and narrative experiences, it becomes important to think about automated ways of dramatic communication as the space of possible variations of individual games increases. A deeper understanding and computational modeling of the way directors and cinematographers craft dramatic experiences will lead to better experiences within games. There are three immediate areas of research that this work leads to. First, incorporating cinematic conventions in a reactive system that interactively works to share the control of the camera with a player in game environments. Second, incorporating a player model and taking into account individual player preferences into the cinematic conventions used by the system. Third, using cinematic techniques to let players record and share their game-playing experiences.

References

1. Arijon, D.: Grammar of the film language. Silman-James Press (1976)
2. Bares, W., Mcdermott, S., Boudreaux, C., Thainimit, S.: Virtual 3d camera composition from frame constraints. In: ACM Multimedia, pp. 177–186 (2000). URL http://citeseerx.ist.psu.edu/viewdoc/summary?doi=10.1.1.94.6163
3. Chatman, S.: Story and discourse: Narrative Structure in Fiction and Film. Cornell University Press, Ithaca, New York
4. Christian, D., Young, R.M.: Comparing cognitive and computational models of narrative structure. In: Proceedings of the 19th National Conference on Artificial Intelligence, pp. 385–390 (2004)
5. Christianson, D.B., Anderson, S.E., wei He, L., Salesin, D., Weld, D.S., Cohen, M.F.: Declarative camera control for automatic cinematography. In: AAAI/IAAI, Vol. 1, pp. 148–155 (1996). URL http://citeseer.ist.psu.edu/christianson96declarative.html
6. Genette, G.: Narrative discourse : an essay in method. Cornell University Press (1980)
7. Graesser, A.C., Lang, K.L., Roberts, R.M.: Question answering in the context of stories. Journal of Experimental Psychology:General **120**(3) (1991)
8. Halper, N., Helbing, R., Strothotte, T.: Managing the trade-off between constraint satisfaction and frame coherence. URL http://citeseerx.ist.psu.edu/viewdoc/summary?doi=10.1.1.12.5345
9. Hornung, A., Lakemeyer, G., Trogemann, G.: An autonomous real-time camera agent for interactive narratives and games. In: Intelligent Virtual Agents (2003)
10. Joseph, M.: The Five C's of Cinematography. Cine/Grafic Publications (1970)
11. Montgomery, D.: Design and Analysis of Experiments. John Wiley and Sons Inc. (2000)
12. Moore, J.D., Paris, C.: Planning text for advisory dialogues: Capturing intentional and rhetorical information. Computational Linguistics **19**(4), 651–694 (1994). URL http://citeseer.ist.psu.edu/moore93planning.html
13. Nau, D., Ghallab, M., Traverso, P.: Automated Planning: Theory and Practice. Morgan-Kauffman Publishers, San Francisco, CA (2004)
14. Polti, G.: Thirty-Six Dramatic Situations. The Writer Inc., Boston, USA. (1921; 1977)
15. Riedl, M., Young, R.M.: Character focused narrative planning. In: International Conference on Virtual Storytelling, Second. Toulouse, France (2003)
16. Sacerdoti, E.D.: The nonlinear nature of plans. In: International Joint Conference on Artificial Intelligence, pp. 206–214 (1975)
17. Sijll, J.V.: Cinematic Storytelling. Michael Wiese Productions, Studio City, CA (2005)
18. Steven, D., David, Z.: Intelligent camera control for graphical environments. In: Graphics Interfaces (1997)
19. Tomlinson, B., Blumberg, B., Nain, D.: Expressive autonomous cinematography for interactive virtual environments. In: C. Sierra, M. Gini, J.S. Rosenschein (eds.) Proceedings of the Fourth International Conference on Autonomous Agents, pp. 317–324. ACM Press, Barcelona, Catalonia, Spain (2000). URL http://citeseer.ist.psu.edu/tomlinson00expressive.html
20. Yorke-Smith, N., Venable, K.B., Rossi, F.: Temporal reasoning with preferences and uncertainty. In: International Joint Conference on Artificial Intelligence, pp. 1385– (2003)
21. Young, R.M., Pollack, M.E., Moore, J.D.: Decomposition and causality in partial-order planning. In: proceedings of International Conference on Artificial Intelligence and Planning Systems. Chicago, IL (1994)

Intelligent Adaptive Lighting

Enhancing the Video Game Experience

Magy Seif El-Nasr, Joseph Zupko, and Chinmay Rao

Abstract The game industry is currently exploring the development of designs that can appeal to a wide market with users exhibiting different tastes, tenancies, behaviors, abilities, and life styles. This problem requires the industry to look for innovative design solutions and tools. Artificial intelligence (AI) techniques can be used to adapt the game experience to players' skills, behaviors, and abilities. Recently, several adaptive systems have been proposed including adaptive character AI [Spronck et al., Online adaptation of game opponent ai in theory and practice. In: 4th International Conference on Intelligent Games and Simulation (GAME-ON 2004), pp. 93–100, 2003] and game design [Charles et al., Player-centred game design: Player modelling and adaptive digital games. In: DIGRA 2005 – Changing Views: Worlds in Play – Electronic Proceedings (2005); Yannakakis and Hallam Real-time game adaptation for optimizing player satisfaction. In: IEEE Transactions on Computational Intelligence and AI in Games, pp. 121–133, 2009]. In this chapter, we discuss an intelligent adaptive system that adapts lighting in a 3D game to enhance the users' experience. Lighting design is well known among designers, directors, and visual artists for its vital role in influencing viewers' perception by evoking moods, directing gaze to important areas (i.e., providing visual focus), and conveying visual tension. The intelligent lighting systems discussed in this chapter adapt lighting qualities, in terms of visual attention and affective properties, by integrating a constraint optimization system built based on cinematic and theatric techniques. The system has been in development and refinement for 9 years. In this chapter, we will discuss the systems as well as their evaluation through different game prototypes, specifically highlighting their effect on the users' experience.

M.S. El-Nasr (✉)
Simon Fraser University, Burnaby, BC, Canada V5A 1S6
e-mail: magy@sfu.ca

P.A. González-Calero and M.A. Gómez-Martín (eds.), *Artificial Intelligence for Computer Games*, DOI 10.1007/978-1-4419-8188-2_8,

1 Introduction

During the past few years, interactive environments that facilitate engagement and involvement, such as social media, 3D games, and social games, have become a very important area of research. The technical contributions and design innovations made to advance such environments have a direct impact on applications used for education, training, entertainment as well as communication and software design. Research within this area was recently acknowledged as a priority area by the Canadian Science, Technology and Innovation Council as well as by the National Science Foundation (NSF) and the National Institutes of Health (NIH) in the USA. This resulted in many funded projects that explore the utility of these environments for health therapy and education, to mention a few applications.

One of the important factors that make these environments popular is their ability to engage users at an emotional level [24, 38]. To stimulate emotional involvement, developers often allocate much time and effort to artistic content development and esthetics, specifically to define the look and feel of the environment. In fact, a typical AAA-title production cycle ranges 2–6 years of production with 100–200 people (artists, programmers, and designers) and over a $20 million dollar budget. While companies differ in terms of the ratio of artists to programmers to designers, most of the companies we have interviewed acquired more artists on their production teams than programmers (personal communication). According to John Buchanan, Technical Director at *Relic Entertainment*, the industry is a content development industry where the development and perfection of art assets and environment design are regarded as the most important factors in producing a successful game. Therefore, visual design and perfecting the look and feel of these environments are considered tasks of highest priority.

Games are interactive, and thus by nature unpredictable, i.e., users are free to do whatever they want in whatever area they want. Currently, however, most games are developed to constrain interactions in ways that can be predictable by designers. This ensures that designers are able to produce the kind of polish needed. While this is true for most games, there are games that rely on procedural content generation, such as open world games, simulation games, etc. For these games, developing procedural encounters and gameplay elements are important. To produce the expected polish procedural visual esthetic systems are required.

In the past, there has been some research in both academia and industry exploring the development of procedural content. For example, *The Sims* (Electronic Arts, 2000) and *The Movies* (Activision, 2005) use intelligent adaptive systems to procedurally simulate emotions and personality within their believable characters. In addition, *Spore* (Electronic Arts, 2009) was also well recognized for its push toward better systems for believable procedural animation. From the academic side, procedural believable characters has been a topic of interest since the 1990s with several researchers making clear and significant contributions, including Mateas and Stern [34], Institute of Creative technology's team at USC [47], Justine Cassell [14], Ken Perlin [42], Seif El-Nasr [22], and the seminal work of the Oz project group at CMU [4, 5, 33]. In addition to the believable character work, the academic community just recently started shifting toward other game procedural systems, including procedural level and environment generation [28,44,45], and procedural game design [40].

As the research on procedural game content start to develop, a need to look into procedural systems for visual esthetics and design will become important. Currently there are very few systems that focus on visual esthetics, such as lighting or camera. Most of the work in graphics have focused on producing a physically correct model, e.g., a physically correct model of lighting [39, 50], or a physically correct model of vegetation [46], etc. Very little work investigated procedural esthetic systems, yet as argued above, a game cannot be successful without the polish and esthetics. Thus, while physically correct procedural modeling is important, developing better procedural visual esthetic systems is even more crucial for emotionally engaging simulations or games.

This chapter focuses on procedural visual design, in particular lighting and its possible effect on the users' experience. Specifically, we will discuss several systems we developed to allow procedural control of lighting while taking visual design goals into consideration. Controlling lighting procedurally within an engaging experience requires consideration of realistic or physical as well as esthetic qualities of light. Unlike the physical qualities, esthetic lighting qualities are not scientifically studied or formalized, and thus are not easily modeled within a procedural system. In this chapter, we discuss the first dynamic esthetic lighting system we developed: the Expressive Lighting Engine (ELE). This system was developed and published in 2003 [19, 21]. Based on this system, we developed two systems: Adaptive Lighting System for Visual Attention (ALVA) and Temporal Dynamic Expressive Lighting Engine (TDELE). Using these two systems, we explore the effect of lighting on two properties of lighting esthetic: evoking emotions or affect and directing visual attention.

The chapter is divided into the following sections. First, we will discuss lighting design from an esthetic viewpoint, detailing the role of lighting within a media production, including theater, film, or interactive media. This will give the reader a clearer picture of the elusive terms used here, e.g., esthetics or lighting design. We will then outline the previous work within the area. Since there are very few research works that target esthetic lighting systems, we will focus on applied techniques used by the game industry. Following this discussion, we will outline three systems that we developed, specifically ELE, ALVA, and TDELE. The last section will discuss some experiments we conducted to measure the impact of ALVA and TDELE on the user experience. The chapter then concludes by discussing the current state and open problems within this area.

2 Lighting Design

> Every light has a job to do, every light must fit and balance within the overall shot, every light interacts with others and with the action. They all work together in a web of complexity.
>
> —Brown 96

Visual esthetics is an elusive term. Since this chapter is concerned with lighting design from an esthetics perspective, we will briefly discuss what we mean by the word "esthetics," specifically lighting esthetics. According to the Stanford philosophy encyclopedia, esthetics is defined as the study of beauty [51]. Specifically, it is the study of emotional sensations that we feel as we see particular designs or artifacts that we deem beautiful based on our own judgments and tastes. Simon Neidenthal [41] offered three different meanings to the term "Game Esthetics." He defined it as a term referring to: (1) "the sensory phenomena that the player encounters in the game (visual, aural, haptic, embodied)," (2) "those aspects of digital games that are shared with other art forms (and thus provides a means of generalizing about art)," or (3) "is an expression of the game experienced as pleasure, emotion, sociability, form giving, etc. (with reference to 'the esthetic experience')" [41]. While these definitions provide a clearer picture of the term, they do not define a workable design model. Thus, in this section, we will use Foss' discussion of narrative functions [25] as a method for defining lighting design functions and lighting esthetics.

When designing lighting for a theater, film, or game production, designers often explore several functions of light; these include *realistic*, *lyrical*, *dramatic*, and *esthetic*. A lighting design achieves realism by conforming to a realistic color palette, conservatively changing light colors to preserve visual continuity, and selecting angles of light that adhere to the direction of practical sources. These decisions increase the credibility of the scene and help the audience identify with and relate to the scene. To achieve lyrical goals, designers need to set up lights to evoke moods or emotions. Lighting designers use several perceptual rules to adjust colors and angles of lights to achieve a desired mood [2, 12]. For instance, lighting designers may vary the degree of visibility of a character's face to affect the audience's emotions and feelings since it is known that less visible faces elicit uneasiness [26]. Another example that is most commonly used is the increase in contrast, i.e., the increase in the amount of darkness within a frame, to elicit fear and a sense of mystery [2, 26]. In addition, lighting is often configured to serve various dramatic goals, including emphasize dramatic tension, attract viewer's attention to important objects or characters, and provide good visibility for the action and characters within the scene. A scene typically follows a dramatic shape [3], which describes the increase/decrease of dramatic tension through time. Lighting designers design lights to parallel such escalation or drop of tension. They use contrast or affinity of saturation, brightness, or warmth/coolness of color to show tension [8]. Even though game designers do not adjust lighting during interaction, they manually select colors to parallel the anticipated tension and mood [13]. For example, in *Silent Hill*, designers used darkness and red-colored tints to signify danger and increase tension when fighting zombies or when zombies are near.

We have conducted a qualitative study using 30 movies, including *The Cook, The Thief, His Wife and Her Lover* (1989), *Equilibrium* (2002), *Shakespeare in Love* (1998), *Citizen Kane* (1941), and *The Matrix* (1999). The goal of the study was to identify cinematic techniques used for lyrical and dramatic functions. According to our study, the techniques used can be divided into shot-based color techniques: color techniques used in one shot, and scene-based color techniques: techniques used on a sequence of shots.

An example for shot-based color technique is the use of high brightness contrast in one shot. High brightness contrast denotes a large difference between brightness in one or two areas in the scene and the rest of the scene. This effect is not new; it was used in paintings during the Baroque era and was termed Chiaroscuro – an Italian word meaning light and dark. This kind of composition has also been used in many movies to increase tension or emotional reaction. Perhaps the most well-known examples of movies that use this kind of effect are film noir movies, e.g., *Citizen Kane* (1941), *The Shanghai Gesture* (1941), *This Gun For Hire* (1942). Another variation on this technique is the contrast between warm and cool colors, which can be seen in many Color Noir movies. These kinds of patterns are usually used in peak moments in an experience, such as a turning point. Lower contrast compositions usually precede these heightened shots, thus developing another form of contrast: contrast between shots.

In addition to color and brightness contrast, filmmakers have also used affinity of color to elicit emotional responses [6, 8–11, 16, 17, 26]. Movies such as *The Cook, The Thief, His Wife and Her Lover* sustain an affinity of highly saturated warm colors for a period of time. The temporal factor is key to the effect of this approach; this is due to the nature of the eye. The eye tries to balance the projected color to achieve white color. Hence, when projected with red color, the eye tries to compensate the red with cyan to achieve white color. This causes eye fatigue, which in turn affects participant's stress level, thus affecting affect. This technique can also be seen in games, such as *Devil May Cry*. In contrast, designers have also used desaturated colors to project low energy scenes, thus decreasing affect. For example, *Equilibrium* (2002) and *The English Patient* (1996) both used low saturation cool colors to increase detachment and decrease affect.

Of course the perception of contrast, saturation, and warmth of color of any shot within a continuous movie depends on colors used in the preceding shots. Several movies used contrast between shots to evoke emotions [2, 8]. For instance, filmmakers used warm saturated colors in one shot, then cool saturated colors in another, thus forming a warm/cool contrast between shots to decrease affect.

In addition to evoking emotions and setting overall moods, lighting designers select angles, colors, and positions for each light in a scene to provide information about the action, the characters, or their relationships. For instance, if the scene's goal is to show distress of character y, then the lighting should emphasize this goal by focusing on character y and adjusting the colors to parallel the negative mood of the character. Also, light angles communicate character traits that emphasize the dramatic motive. For example, a character that is under-lit is often characterized as sinister or mysterious [12].

Lighting designers also consider modeling and depth that can be created and stimulated through lighting in a scene [6]. A widely used technique for modeling is 3-point lighting, where three lights are used to light a character: a backlight, a key light, and a fill light. Backlight is positioned behind the character with a slight downward angle and is used to separate the character from the background. Key light provides the key source of illumination and is positioned at a front offset angle to emphasize texture and shape. Fill light is positioned to mirror the effect of the key light. Another important esthetic quality of light is depth. Depth is established by varying the colors and contrast between lights lighting the background and those lighting the foreground of a scene.

The role of a lighting designer is to establish a lighting design that serves the goals described above. However, these goals are often in conflict with one another. To illustrate this problem, consider the following example. A scene is established in a room with one window. The lighting designer adjusts the angles of light to appear as if coming from the window. At some point the character stands in a corner that is not directly lit by the window. According to the director, this moment is very important for the dramatic development of the scene, and the character's face should be sufficiently visible. Therefore, to achieve the dramatic goal, the lighting designer should add lights to establish good visibility. On the other hand, if the lighting design is to serve realism, lights that do not conform to the direction of light emitted by practical sources should not be added – and there lies the lighting designer's dilemma.

This example illustrates how lighting design, as many design tasks, involves tradeoffs. Lighting designers favor some goals over others depending on the lighting style chosen and the dramatic situation. In the example above, for instance, if the lighting designer chooses a realistic style, he or she will sacrifice the dramatic goal. On the other hand, if he or she chooses to conform to the dramatic and realistic goals, then character blocking and camera placement will need to be changed.

3 Current Game Lighting Techniques

The lighting design process in games is deeply tied to the rapidly changing rendering technology. The game industry has developed several approaches to lighting. Until recently, most video game lighting was precalculated (according to an interview in [52]). Often, an algorithm such as radiosity [35] is used and the radiance[1] of a surface is "baked" into the surface using light maps, which store the view-independent radiance of the surface. This approach supplies lighting designers with the freedom to allocate a lighting setup that fulfills dramatic goals set by the static story or setup [6, 11].

With current technologies and the emphasis on developing more responsive games considering the interaction, there is a move toward incorporating dynamic lighting (interview presented in [52]). The dynamic lighting approach is based on mathematically simple light primitives and is calculated at runtime. Real-time rendering typically allows only a few visible dynamic lights per object (8–10), but

[1] Radiance is the light emitted by a surface.

deferred rendering technologies [31] allow for many visible dynamic lights (40–50). However, due to tradeoffs with a deferred approach [37], non-deferred rendering is still often used; hence, it is not safe to assume that more than a few dynamic lights can be used per object in a real-time rendering environment.

Still much of the industry techniques rely on static lighting. However, technological advancements in rendering allows designers to display better lighting, using spherical harmonics [27]. Spherical harmonics allow irradiance[2] to be stored at points in space and efficiently applied at runtime. Samples are commonly stored as either a surface-aligned texture similar to light maps or in coarsely spaced grids [15]. Although spherical harmonics require more storage and processing than light maps, they offer the capacity to produce effects that light maps cannot, such as bump mapping [7] and low frequency specularity [15]. Spherical harmonics are mostly static but can be rotated [27].

Although this kind of static approach is highly restrictive as it cannot dynamically adjust based on interactions or context, it is the approach that is most widely used in the industry. The reason is that in most games it is often desirable to constrain the interaction to provide a more manageable and predictable game; i.e., designers often constrain the interaction to specific spaces and develop systems that are predictable allowing lighting designers to manually light the space, given the camera perspective.

Even though the constrained and predictable approach is the most widely used, there are opportunities to develop and incorporate procedural lighting, but this will highly depend on the game genre. For example, games like open world games or procedural games would benefit from procedural content and procedural visual design. In addition, as technology advances and games start to target different markets, adaptive systems will become a necessity. Thus, procedural visual design systems may at some point be incorporated into different genre depending on how technology and designs progress.

4 Current Lighting Systems

In the past few years, we have experimented with several ideas for adaptive lighting design tools that can be used to enhance the game experience. There were two goals to these lighting systems: (1) a clear impact on the gaming experience, i.e., these systems enabled a better gaming experience, and (2) a clear impact on the utility of these systems as tools for designers in a production process, e.g., cutting down production time or enabling better or faster prototyping at preproduction. In this chapter, we will only focus on (1).

[2] Irradiance is light incident to a surface.

4.1 Expressive Lighting Engine

ELE was developed and published in 2003 with the two goals above in mind [18,21]. Since the goal was to develop a tool that will allow designers to light a scene within a game or interactive narrative environment, and since we know that esthetic, contrast, and the look and feel of an environment is of great importance to designers, it was required that the lighting system accommodate esthetic goals. Previous work on lighting, as discussed above, mostly concentrated on physically based lighting with no focus on esthetics. Therefore, creating a model for lighting design became an important research goal, especially since there are no esthetic lighting models that can be used to derive the system. Additionally, since game environments are normally interactive, and thus unpredictable, the lighting system will also need to adjust the lighting dynamically in real-time given specific contexts or based on specific constraints that the artists or designers indicate. Therefore, ELE was developed with the following two main requirements in mind: (1) adapt to the unpredictable environment and (2) provide a method for manipulating and encoding esthetic lighting design. Discussing the esthetic model behind ELE and how ELE achieves dynamic lighting is beyond the scope of this chapter; readers are referred to [18, 21] for more details. However, we will summarize the system here.

ELE uses constraint optimization algorithms to compose and adapt lighting design dynamically and in real-time, accommodating user interaction while achieving esthetic design goals. ELE automatically, in real-time, selects and modulates the lighting configuration, including number of lights, their positions, colors, and angles, satisfying several visual design (or perceptual) goals: directing visual attention, establishing depth, accommodating visibility, evoking moods, paralleling dramatic tension, and maintaining visual continuity.

These design goals were identified based on cinematic and theatric lighting design theories as well as tacit knowledge collected by the first author while working in theater lighting. These goals are entered as numeric constraints into ELE. It should be noted that these goals also conflict with one another. For example, choosing a lighting configuration that establishes a particular atmosphere or mood may hinder visibility. ELE takes that into account.

ELE is composed of three subsystems, shown in Fig. 1, a lighting allocation subsystem selects best number of lights and placements, an angle subsystem selects best angles for each light, and a color subsystem selects best color for each light. ELE sits on top of a rendering or game engine. All movements and changes in the environment are communicated to ELE using an xml structure called *WAMP*. Similarly, ELE communicates the layout, angles, and colors for each light in another xml format called *LAMP*, which is then passed on to the rendering engine.

The layout subsystem uses level layout, scene graph information, and artistic constraints to create a light layout. The system divides the level into n different areas and then categorizes these areas as focus, non-focus, or background. This information is then used to achieve the design goal mentioned above. To compute the

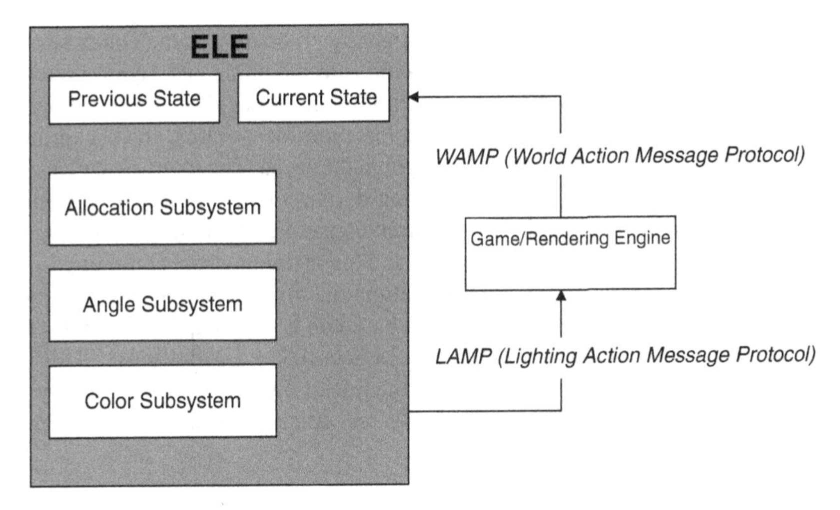

Fig. 1 ELE's architecture

light layout based on these goals, the system minimizes a multi-objective function to determine the number of lights to use for each area:

$$p_{opt} = \arg\max_{p}[\lambda_v V(p) + \lambda_d D(p) + \lambda_m M(p) + \lambda_{VC} VC(p)]$$

where p is the light configuration and λ are weights representing constraints: λ_v is the importance of visibility, λ_d is the importance of depth, λ_m is the importance of modeling, and λ_{vc} is the importance of visual continuity. $V(p)$ is visibility given p, $D(p)$ is depth given p, $M(p)$ is modeling given p, and $VC(p)$ is visual continuity given p.

In determining the angles of light, ELE also takes into account the use of light angles in projecting depth, modeling, and mood, where mood is evoked through the angle of light on a character. For example, a character can be lit from below, creating a sense of evil or mystery. ELE uses nonlinear optimization to select an angle for each key light that minimizes the following function:

$$\lambda_v[1 - V(K,s)] + \lambda_{-}|k - k^{-}| + \lambda_m|k - m| + \lambda_l \min_{i} |k - l_i|$$

where k and s are defined as the key light azimuth angle relative to the camera and the subject angle relative to the key light. k^{-} is the key light azimuth angle from the previous frame and the λ_s represent artistic constraints. Specifically, λ_{-} is the cost of changing the key light angle over time (to enforce visual continuity), λ_m is the cost of deviation from the mood azimuth angle, m is the mood azimuth angle suggested by the artist, λ_s is the cost of azimuth angle deviation from a practical source direction, l_i is the azimuth angle of light emitted by the practical source i, and λ_v is the cost of deviation from an orientation of light that establishes best visibility.

As mentioned above, the interaction of lighting colors in a scene composes the contrast and feeling of the entire image. Similar to the angle and layout systems, ELE uses a nonlinear optimization to search through a nine-dimensional space of RGB values. It differentiates among focus colors, non-focus colors, and background areas to select a color for each individual light in the scene. The multi-objective cost function evaluates a color against the lighting design goals, including establishing depth, conforming to color-style and constraints, paralleling dramatic tension, adhering to desired hue, saturation, lightness, and maintaining visual continuity.

ELE was the beginning of several explorations that followed. In Sect. 4.2, we will discuss ALVA which is a system developed based on ELE to explore the impact of manipulating lighting within an FPS game. In Sect. 4.3, we will discuss TDELE a system based on ELE developed to explore the role of lighting in evoking emotions. Section 5 will discuss experiments we conducted with these systems showing their impact on the user experience.

4.2 Adaptive Lighting System for Visual Attention

While ELE included a goal for manipulating visual focus, it did that through adjusting brightness of the focus area in comparison to the non-focus area. The problem with this approach is that it does not always generate the right results, especially when the non-focus areas are lit with very bright light. Instead, we developed a new system ALVA that encodes several rules based on psychology of visual attention [49], thus extending ELE's ability to manipulate the lighting specifically for visual focus.

Figure 2 shows the architecture for ALVA. We indicated the places (shaded in the figure) where ALVA adds a significant improvement to ELE. We will concentrate on these main shaded areas in this section.

4.2.1 Layout

To achieve and maintain visual attention, ALVA uses some game parameters, including level or zone configuration, quests, number of characters, number of objects within the scene, architectural points of importance, their dimensions, camera position, and anticipated movements. It uses these parameters to identify areas of visual focus. Unlike ELE, ALVA identifies these focus areas using authored rules that designers encode in the system. These rules identify when specific objects or characters become important and how important they are given a game state, player's goals, and missions. An example rule can be as follows:

```
(defrule
  trigger: (goal ?player (get key1))
  action: (attend-to questobjectID123 100))
```

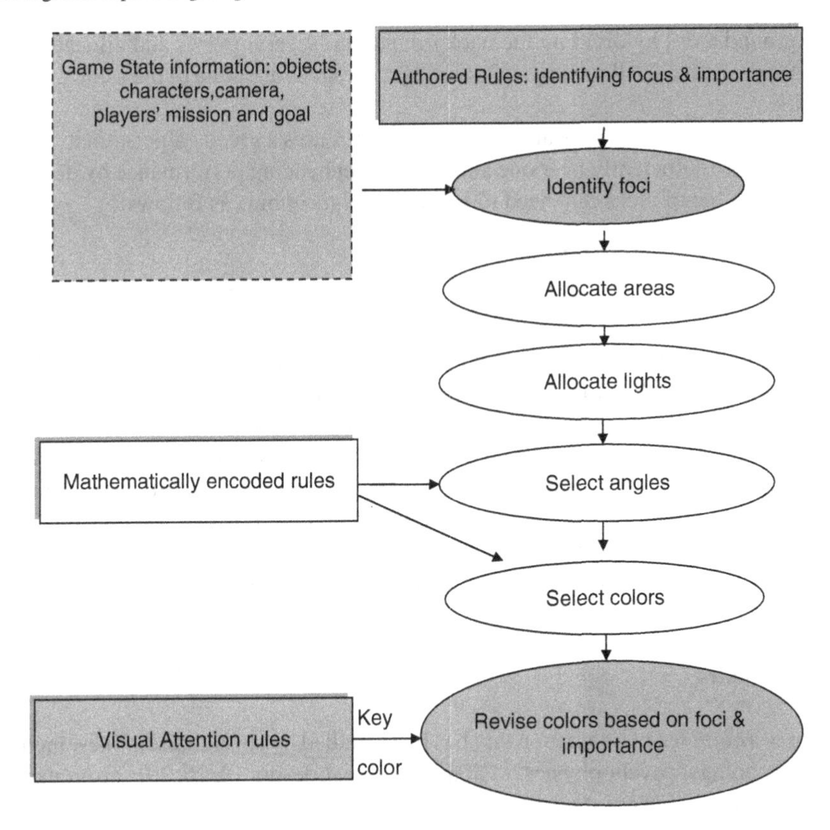

Fig. 2 ALVA's architecture. *Shaded areas* signify areas of enhancements on ELE

where `questobjectID123` is a quest object related to the goal of getting `key1`. Thus, this rule indicates that if the player's goal is to get `key1`, then the `questobjectID123` should be highlighted with 100% importance. Examples of important objects ALVA can highlight include, visible enemies, visible quest objects, and parts of a level identified by the designer as an object that can lead the player forward (this is important for platform games or spatial puzzles games like *Prince of Persia*).

Based on these authored rules and the game state, ALVA creates focus areas, which are represented as cylinders around the important objects. Unlike ELE, ALVA assumes several areas of focus rather than just one focus area. It also gives each area an importance level. This is important for games, because most often there are several areas that are important for the players to attend to with different attention levels. ALVA first determines these focus points and their importance.

Like ELE, ALVA divides the visible area into several foreground areas depending on the maximum number of lights that can be used, the number of non-focus objects in the level, and the focus objects computed. Additionally, ALVA creates several background areas by dividing the background into several pieces and allocating an area for each piece. Allocating and dividing these areas are done in the same manner as ELE.

Once these areas are allocated, like ELE, ALVA uses a greedy algorithm to merge areas that are sufficiently near one another, thus enhancing performance by decreasing the number of areas that need to be lit. The algorithm is as follows:

```
Repeat for each area a
if ∃a′ s.t. |a − a′| < ε, and both are focus areas with
same importance (or non-focus) then merge a, a′
```

It then allocates some lights to each area. Once lights are allocated, ALVA assigns angles and colors to each light such that visual foci are established and visual continuity is maintained. Since colors impact visual attention, we will discuss colors here. Readers can assume angles to be set to 45° azimuth and 45° elevation relative to characters' faces, which are good angles for establishing visibility and character modeling as defined by [36]. It should be noted that the process of allocating light areas, angles, and colors occurs on each event, including change of camera angle, change of location of an object, on entrance of a level, etc.

4.2.2 Colors

Color is a complex phenomenon that has been studied by several disciplines, including psychology, psychophysics, vision, and visual design [8, 26, 29]. From earlier work, we know that color affects attention and emotions [8, 11, 35]. These research works have identified several features of color that play a significant role in affecting attention: contrast, warmth, and brightness.

To use these features, we need to first define the attributes of color and develop formulae for defining these attributes in terms of RGB values. We define the attributes of color that are of interest as: brightness, warmth, and saturation. We calculate lightness instead of brightness. Lightness and saturation are calculated by transforming the RGB color to the HSL color space [43].

Warmth, on the other hand, is an elusive quality. It impacts our attention, as discussed by Block and Treisman [8, 48]. Warm colors are defined to be colors with high proportion of reds and greens [29], while cold colors are colors with high proportion of blues relative to the reds and greens [29].

Several psychology and psychophysics theories describe warm and cool colors. However, none of them presented data that can be used to formulate warmth and coolness of colors in terms of the HSL or RGB color models. The best effort to measure this elusive quality is described in an unpublished paper by Katra and Wooten. They gathered results from several experiments in which subjects rated colors on a scale of -5 to 5, where 5 is warm and -5 is cold. The stimuli were controlled for hue and saturation [30].

Based on their results, we used a multiple linear regression method to formulate an equation describing color warmth, described in RGB color space. The formula is as follows:

$$\text{warmth}\left(\begin{bmatrix} R \\ G \\ B \end{bmatrix}\right) = \begin{bmatrix} 0.008 \\ 0.0006 \\ -0.0105 \end{bmatrix}^T \begin{bmatrix} R \\ G \\ B \end{bmatrix} - k$$

where k is a constant.

Another important concept that is constantly used by ALVA is contrast. Film and theater lighting designers differentiate between three different types of contrast: lightness, warmth, and saturation contrast [8]. Contrast is measured as the difference between the lightness, warmth, or saturation of lights lighting the focus areas compared to lights lighting the surrounding areas. We use the same formulation used by ELE to measure contrast:

$$\text{contrast}_\phi(c) = \sum_{i \neq \text{focus}} |\phi(c_{\text{focus}}) - \phi(c_i)|$$

where ϕ represents either lightness, warmth, or saturation.

4.2.3 Contrast and Balance

ALVA first assigns colors to each light in the level in a similar fashion to ELE, where constraint-based optimization is used to choose best colors for all lights in the scene to accommodate artists' constraints based on esthetic color choices, mood, and depth. Once this is done, ALVA then revises the color assignments based on visual attention, as shown in Fig. 2. Thus, the assignment of color happens in two phases. This was done to simplify the equations involved for constraint optimization and provide modularity.

ALVA revises the color assignments based on visual attention. Research showed that our attention is directed toward warmer colored objects when they are surrounded by cool colored objects, but the impact may not be the same if a warm colored object is surrounded by objects whose color projects the same degree of warmth [8, 29, 49]. Thus, contrast is key to modulating and adapting visual focus.

Following this theory, ALVA embeds several rules to manipulate colors of lights on the objects. These rules fall into several cases: level entry, changes in positions within the level, and during an event that stimulates a lighting change. It is important to differentiate between these cases because the constraints imposed on lighting changes are different for each case. For example, upon an entry of a level, there are no constraints on lighting changes with regards to visual continuity. This is because the player has never seen the level before, and so it is fine to perform any kind of edits. During play within the level, lighting should incur very little changes, especially overall level lighting due to the desirability to maintain visual continuity. If there is a lighting motivation, i.e., there is an event that causes lighting change, then the system has freedom to change the lighting connected to the event.

ALVA is a beginning for us to explore the impact of lighting on perception and user experience. In Sect. 5, we will discuss some of the experiments we ran to determine the impact of using ALVA or a system like ALVA on user experience.

4.3 Temporal Dynamic Expressive Lighting Engine

We have adapted ELE to embed patterns of lighting that evoke tension based on our previous qualitative study [1,20]. We will briefly review these patterns here; the reader is referred to [1] for more details on the films and games used as well as the method used to extract these patterns. Based on the qualitative study discussed in Sect. 2, we formulated the following patterns:

Pattern I Subjecting audience to affinity of high saturated colors (where high saturation ranges from 70% to 100%) for some time increases arousal.

Pattern II Subjecting audience to contrast in terms of high saturated then low saturated colors (where saturation ranges from 100% to 10%) over a sequence of shots decrease arousal.

Pattern III Subjecting audience to contrast in terms of low saturated then high saturated colors (where saturation ranges from 10% to 100%) over a sequence of shots increase arousal.

Pattern IV Subjecting audience to contrast in terms of high brightness then low brightness (where brightness ranges from 100% to 10%) over a sequence of shots decrease arousal.

Pattern V Subjecting audience to contrast in terms of low brightness then high brightness (where brightness ranges from 10% to 100%) over a sequence of shots increase arousal.

Pattern VI Subjecting audience to contrast in terms of warmth then cool colors (where warmth ranges from 100% to 10%) over a sequence of shots decrease arousal.

Pattern VII Subjecting audience to contrast in terms of cool then warm colors (where warmth ranges from 10% to 100%) over a sequence of shots increase arousal.

Pattern VIII Subjecting audience to increase of brightness contrast subjected in a shot (where brightness contrast is measured in terms of difference between bright and dark spots in an image) over a sequence of shots increases arousal.

Pattern IX Subjecting audience to decrease of brightness contrast subjected in a shot (where brightness contrast is measured in terms of difference between bright and dark spots in an image) over a sequence of shots decrease arousal.

Pattern X Subjecting audience to increase of warmth/cool color contrast subjected in a shot (where contrast is measured in terms of difference between warm and cool spots in an image) over a sequence of shots increases arousal.

Pattern XI Subjecting audience to decrease of warmth/cool color contrast subjected in a shot (where contrast is measured in terms of difference between warm and cool spots in an image) over a sequence of shots decreases arousal.

Based on these patterns, we adapted ELE and embedded these patterns developing a new system called TDELE. TDELE extends ELE by adding a state that keeps track of ticks (simulation time) as well as the history of lighting color compositions used in the past. This state is represented as a list of light colors for each area as well as contrast value and contrast type; color values are stored in terms of RGB and HSL as well as calculated warmth value. Based on this state information, the desired pattern given the patterns listed above, and the desired tension level, the system calculates constraint values, including desired saturation level, desired warmth value, and desired contrast level. These values are then given to the system to manipulate the current frame. Note that ELE already balances these values with the required visibility, motivation, etc. Therefore, the resulting lighting setup created presents a balanced lighting design.

There are several advantages to using such a system. First, the system embeds several patterns that are not used in the current dynamic lighting design methods in games. Second, it presents a system that establishes a well-balanced lighting design. Third, it allows designers to quickly compose the scene by just choosing the pattern and tweaking it, rather than redesigning the lighting in every level.

We created two prototypes using TDELE. The first prototype was developed to evaluate the use of the patterns to evoke affect. For this purpose, we created an interactive 3D environment using WildTangent, a publicly available web-based game engine; the environment is shown in Fig. 3. The task of the user was to navigate through the environment. The lighting conditions were varied as a function of time. The figure shows three screenshots taken at different times during the interaction. The lighting system was configured to use pattern IX, where brightness contrast subjected was decreased as a function of time (where brightness contrast is measured in terms of difference between bright and dark spots in an image). As the figure shows, visibility was well balanced with the contrast effect, as contrast increases or decreases in time.

The second prototype is a First Person Shooter (FPS) game that uses an implementation of TDELE on top of the Unreal Engine [23]. An interface was added to enable designers to integrate their own tension formula and link it to these patterns. For example, they can define tension as the rise and fall of health within a FPS game. In this case, the TDELE manipulates the lighting in the room to project rise and fall of health and number of enemies as a symbol of tension. It should be

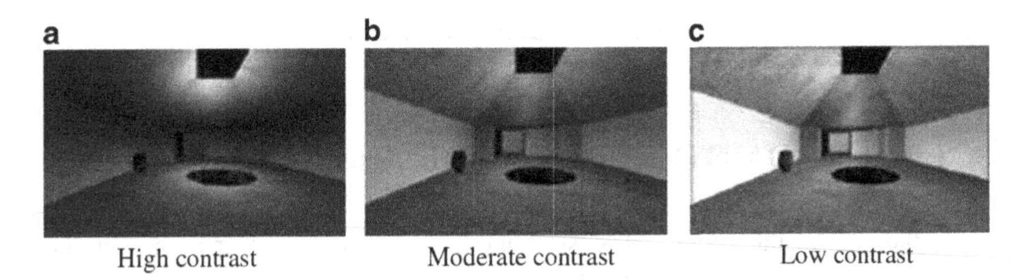

Fig. 3 Linearly increasing brightness contrast (where center of room is the focus)

Fig. 4 Varying brightness contrast

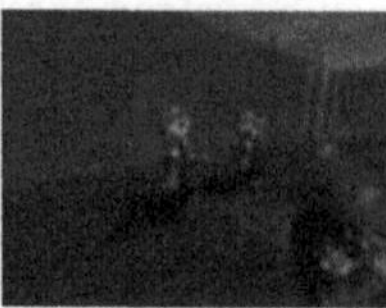

Fig. 5 Linearly increasing saturation

noted that designers can use this tool to induce any of the patterns discussed above. Thus, they can project the same lighting effects, such as the use of saturated warm colored lights, etc. They can also use it to create a contrast effect such as high contrast (e.g., leftmost screenshot in Fig. 3) and sustain it over time.

The lighting compositions varied with the level composed. For example, in the beginning a decrease of brightness contrast was established through the opening scene, shown in Fig. 4. During the game, the designer authored rules within TDELE to increase or decrease tension based on tension as defined above. The author used a combination of patterns I and III, where increase in tension was projected as an increase in warmth and saturation of surrounding lights. Thus, if the user is confronted with many monsters and his or her health is dropping over time, the warmth and saturation of color will increase over time showing an increase in tension. While if the player is killing monsters and danger level is diminishing, warmth and saturation will decrease through time. Screenshots from the game are depicted in Fig. 5. A video of the demo can be found at URL: `http://www.sfu.ca/~magy/ondisplay.html`.

5 Effect of ADAPTIVE Lighting on the Users' Experience

Since the goal of the systems discussed above was to have a clear impact on the gaming experience, we developed studies that attempted to verify and validate these goals. Results of these studies will be discussed in this section.

5.1 Results of Using ALVA in an FPS Game

To measure visual attention within a 3D game and compare the use of lighting for visual attention, we designed an experiment where we asked 26 students from a 300-level undergraduate class at Penn State University to play two games: (1) a game mod of Unreal we developed with static lighting and (2) the same game mod of Unreal but with ALVA dynamically adapting the lighting for best visual attention. The order by which the games were introduced was chosen randomly to balance the order effect. Additionally, we asked participants to wear an eye tracker. We recorded their eye movements superimposed on the game video for later analysis. We also observed and analyzed their behaviors to determine points of frustration and engagement. At the end of each session, we interviewed them to gauge their experience and engagement.

From the 26 students, only 16 were usable due to problems with the eye tracker data. Ten of the 26 subjects wore contact lenses, or had dark eye colors or dark eye lashes, which caused calibration problems. From the 16 students, 13 students identified themselves as non-FPS gamers: three non-gamers and ten casual games, while three identified themselves as FPS gamers. We asked them to sign up for a 30-min session. The experimenter introduced the procedure and then asked them to wear a head-mounted eye tracker to track their gaze locations. For this experiment, we used ISCN ETL-500 eye tracker. The experimenter asked them to play *Soul Calibur II* on the Play Station for 10 min. During this time, he or she calibrated the eye tracker. Also, this 10-min session was used to get them acquainted with wearing the eye tracker while playing.

After this 10-min session, we asked them to play the two Unreal Tournament games, for 10 min each. The game developed for this experiment was composed of seven different environments (levels). The objective was to get to the exit going through all seven environments without dying. We asked them to play the game for 7–10 min. If they die, they were asked to restart the game. A normal play through the game took an average FPS gamer 7 min to complete. We asked them to play the two different versions of the game for 10 min each: one with ALVA and the other with static lighting. Figure 6 shows two screenshots from a level within the game; screenshot shown on the right shows the level with ALVA, and the one on the left shows the same point in the level but without ALVA. As can be seen the character in the figure to the right is more noticeable due to the obvious light glowing around him than the character on the left screenshot. It should be noted that no specific treatments were made for the lighting embedded in the enemy textures.

Fig. 6 Two screenshots comparing the system with ALVA (*right*) and without ALVA (*left*)

While playing the game, we asked participants to shout when they see an enemy. The order of which version of the game they played first was randomized to minimize the order bias effect. After each session, we briefly interviewed them asking them to reflect on their experience.

During the play sessions, two researchers took observational notes identifying points of frustration, engagement, searching and environment scanning moments, and enemy misses. We also recorded their game play sessions; the video recorded showed their eye movements as a white cursor superimposed on the actual game play video. We used a counter to keep track of simulation time, measured in *ct* (counter time) rather than in seconds. We noted every time the user shouted that he or she spotted an enemy. We noted places when subjects did not remember to shout and were reminded to do so.

Using this video, we were able to analyze the time, in terms of simulation counter time, it took them to spot an enemy, given the time the enemy appeared and the time they shouted that they spotted him. In analyzing the taped and observed interaction, we noted all enemy clear misses, where participants visually scanned the environment and missed the enemy even though he or she was not hidden in the shadows or behind obstacles. We also noted the number of times the player was killed during the play session.

By analyzing the observed and videotaped responses, we deduced that non-gamers and casual gamers were able to spot the enemy faster with the dynamic lighting system than without it, and thus were able to survive longer (see Fig. 7). Using ALVA, we calculated the average time for spotting enemies as: 4 *ct* for gamers, 26 *ct* for casual gamers, and 46 *ct* for non-gamers. Using the static lighting system, average times were 8 *ct* for gamers and 38 *ct* for casual gamers. For non-gamers, it was harder to quantify the spotting time due to the fact that most of the time they did not spot an enemy before they were killed. Please note that the numbers above were deduced manually and may not be accurate figures due to the eye tracker time lags, time it took us to confirm a spotting event, and the movement of the player. However, since the same procedure was done on data collected from gamers as well as non-gamers, we deduce that the difference presented is valid.

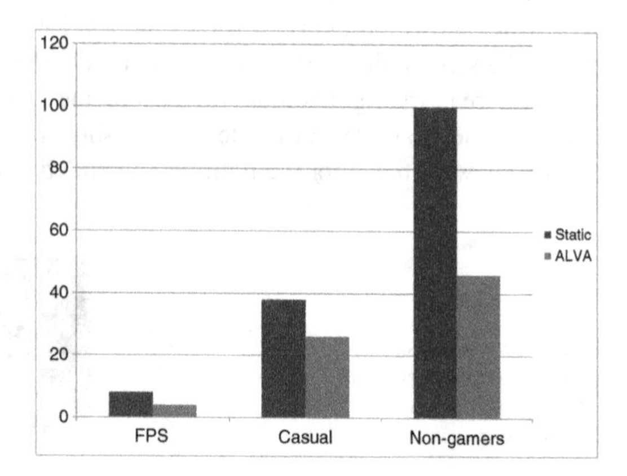

Fig. 7 Difference in counter time of spotting times between FPS gamers, casual gamers, and non-gamers

Table 1 *t*-Test analysis on results of counter time of spotting

	Mean	SD	*t*-Test	Significance
FPS gamers-Static	8	2.65	2.449	95% confidence, no statistical significance
FPS gamers-ALVA	4	1		
Casual gamers-Static	38	13.17	2.2832	95% confidence, statistically significant
Casual gamers-ALVA	28	10.14		
Non gamers-Static	100	0	15.59	95% confidence, statistically significant
Non gamers-ALVA	46	6		

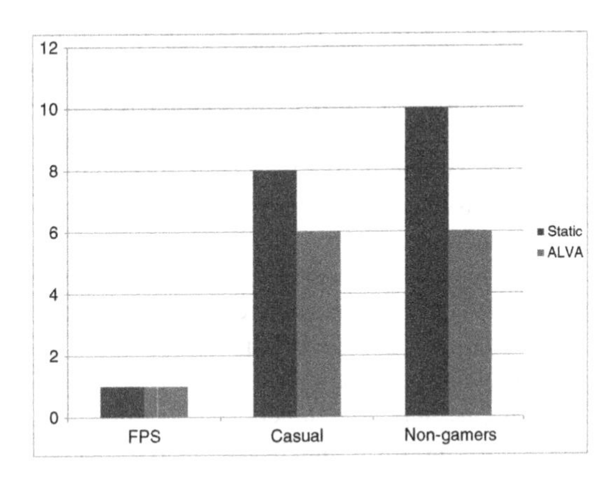

Fig. 8 Difference in deaths times between FPS gamers, casual gamers, and non-gamers

We ran a *t*-test analysis on this data comparing each group and deducing significance. Results are shown in Table 1. As it can be seen, for FPS gamers, their spotting counter time did not show any significance with or without ALVA. But for casual gamers and even more for non-gamers, the results of spotting time show significant improvement with ALVA than without. This leads us to conclude that the introduction of ALVA significantly improved the enemy spotting time of casual gamers and non-gamers. It should be noted, however, that the number of participants are low to draw any conclusive evidence. However, it is still worthwhile to note these results and variance to show the success of the method and suggest an opportunity for further exploration.

The number of times players died before reaching the end of the level varied. In comparison with the static lighting approach, the number of deaths for casual gamers and non-gamers was considerably less (Fig. 8). The figure shows the maximum deaths that occurred over all categories, rather than average. As with the game with static lighting, FPS gamers' death times was 0–1, where two gamers achieved the objective in 7 min with no death and one achieved the objective in 10 min with one death. For casual gamers, death times varied from two to six, with six of them completing the level within the allotted 10 min and four were unable to complete the level. One of the non-gamers completed the level; the number of deaths ranged from 4 to 6. None of the non-gamers quit before our imposed deadline of 10 min.

We also ran a *t*-test analysis on this data comparing each group and deducing significance. Results are shown in Table 2. Similar to spotting time, the number of

Table 2 t-Test analysis on results of number of deaths

	Mean	SD	t-Test	Significance
FPS gamers-Static	0.33	0.58	0	95% confidence, no statistical significance
FPS gamers-ALVA	0.33	0.58		
Casual gamers-Static	6.7	2.214	2.6364	95% confidence, statistically significant
Casual gamers-ALVA	4.4	1.65		
Non gamers-Static	9.33	1.15	4.914	95% confidence, statistically significant
Non gamers-ALVA	5	1		

deaths of FPS gamers was not significantly different with the system with ALVA and the system without. However, for casual and non-gamers the results show statistical significance in the reduction of the number of deaths of these players when playing within the system with ALVA than the system without. This leads us to conclude that the introduction of ALVA significantly improved the number of deaths of casual gamers and non-gamers.

Interviews with players emphasized that gamers were fine with the game, non-gamers and casual gamers were happy with the game, but some expressed inability to react quickly enough to win. It should be noted that playing an FPS requires fast reflexes in addition to fast spotting times. Thus, these results are not surprising. However, helping non-gamers spot enemies faster achieved slightly better results and less overall frustration based on our observations.

5.2 *Results of Studies Exploring the TDELE*

5.2.1 Results from Simple Environment (Prototype 1 Shown in Fig. 3)

To validate the effect of lighting on affect, we conducted several experiments testing each pattern individually within a 3D interactive environment using the simple environment shown in Fig. 3. The environment was composed of six rooms where participants are only allowed to navigate through the environment; thus no object manipulation was required [1].

To validate whether there was an influence on arousal with time, we monitored participant's physiological responses using the SenseWear® PRO_2 Armband, a wearable body monitor that enables continuous collection of low-level physiological factors. It includes several sensors that continuously gather heat flux, skin temperature, near body temperature, and galvanic skin response data from the body. However, it does not include a heart rate monitor. We used the Triax Elite for that purpose. It is composed of a stopwatch and a heart rate monitor strap that displays heart rate, current running pace, and pace target information for interval training. We were able to collect readings from these devices and feed it through MATLAB for further data analysis.

We ran 19 experiments, one experiment for each pattern identified. For each experiment we gathered readings from 20 to 24 participants. We asked participants to volunteer for more than one experiment if they can. If they signed up for more than one experiment, we asked that when they sign up for times to perform the experiment such that experiments are 1 week apart. We think a 1-week period is enough time to eliminate bias of prior exposure to one experiment on the results collected in the other experiment. Participants were male graduate and undergraduate students between the ages of 18–30 from the Information Science and Technology, Electrical Engineering, and Computer Science departments of The Pennsylvania State University. Students were recruited from classes taught in Computer Science, Information Sciences and Technology and Electrical Engineering. We specifically targeted male students due to fluctuations on female hormonal state which may affect arousal, and thus interfere with the results.

All 19 experiments followed the same procedure. In accordance to IRB regulations, before they started the experiment, participants were given consent forms and a brief introduction of the entire experiment. We also asked them to take a color blindness test to check whether they can differentiate between colors. Data collected from participants who did not pass the color blindness test were ignored from the data analysis phase; however, they were allowed to continue the experiment and were given credit. Since the experiment is based on color, we require participants to be able to differentiate between colors.

Once they are done with the test, participants were asked to wear the *BodyMedia* device around their arm and the heart rate device around their chest. They were then asked to navigate within a 3D environment that does not exhibit any of the patterns depicted above. This was done to allow them to relax and get acquainted with the controls and the environment.

After 1–2 min of interacting within this environment, we asked them to navigate through one of the 19 environments developed for this study. Later, participants were asked to fill out a questionnaire which was used as a self report. This procedure was repeated for all the patterns discussed above. We used a different pool of people for each pattern to alleviate the bias of knowing the environment. The experiments were conducted over a period of 3 weeks.

All data collected through the physiological devices were analyzed using a linear prediction (LP) model with the covariance method. In the LP model, also known as the autoregressive (AR) model, the current sample $x(n)$ is approximated by a linear combination of past samples of the input signal [32]. We are then looking for a vector a, of d coefficients, d being the order of the LP model. Provided that the a vector is estimated, the predicted value is computed simply by FIR filtering of the p past samples with the coefficients using equation:

$$\hat{x}(n) = \sum_{i=1}^{p} a_i x(n-i)$$

We used the covariance method of least squares LP to design a casual linear estimator for an output or target sequence based on an input (evocative) sequence. As shown above, most patterns describe a linear relationship between color properties over time such as warmth, contrast, or saturation. Thus, a linear estimation model will suffice. We built an estimation filter using the response data recorded from participants.

The participants were divided into a test and a training set of equal number. The physiological response of the training set and the input color pattern are used to design a filter that is applied on the test set physiological response. The $p+1$ response of the test set is estimated and compared to the recorded $p+1$ value of the response. This approach was used to test the hypothesis of whether individual visual parameters affect physiological response variables.

If the mean-squared error on a test set is sizeably smaller than the variance of the output sequence for some amount of delay/latency between the causal inputs and the target output, it indicates that there exists a linear correlation between the two sequences. If it is not, then a linear correlation does not exist. Using this method, we can deduce with some confidence that there is a linear relationship between the participants' arousal and the pattern used. All results show that the variance of the color patterns was more than the error test values, indicating that there is a linear correlation. Due to space limitations, we will not discuss results of all 19 experiments here, but instead will discuss one set of experiments. Readers are referred to [1] for more details.

We ran subjects through six environments set up with high saturation with six different colors (red, green, yellow, orange, blue, and cyan). The hypothesis is that arousal will increase as a function of time within a high saturated environment. Results shown in Table 3 confirm this result. Specifically, using the data analysis approach, we derived the results depicted in Table 3. As shown in the table, the variance of the color patterns was more than the error test values; this indicates that there is a linear correlation between the expected response (linear increase in arousal

Table 3 Physiological data analysis of red environment

		GSR	Heat flux	Temperature
Red	Variance	2.40E-04	6.963	0.0207
	Error test	7.19E-05	4.1778	0.0124
Yellow	Variance	1.92E-06	1.1147	4.66E-04
	Error test	1.15E-06	0.8917	3.72E-04
Orange	Variance	6.93E-04	4.1521	0.0068
	Error test	4.16E-04	2.4913	0.0041
Cyan	Variance	2.23E-06	0.327	5.36E-04
	Error test	8.91E-07	0.0473	2.14E-04
Blue	Variance	4.94E-06	0.6724	0.0021
	Error test	3.46E-06	0.4707	0.0014
Green	Variance	1.30E-05	3.0224	0.0039
	Error test	9.08E-06	2.1157	0.0027

through time) and the physiological response. The low error rate also indicates that all participants had the same reaction, an increase in arousal as time passes within a 100% saturation. As shown by the results, the specific colors do not have any significant impact on arousal.

These experiments and results validate the relationship between arousal increase/decrease as a result of manipulation of lighting and texture colors over time. However, we did not attempt to deduce valence component of emotions within this experiment. An interesting future direction is to explore the relationship between these patterns and valence or emotions. We anticipate that these patterns will have specific influence on arousal and valence but may not induce specific emotional states, but still, this needs to be validated.

5.2.2 Results of Using TDELE Within a Game

As discussed in Sect. 4, we also developed a game as a mod to the Unreal environment that exhibits the patterns defined in TDELE as we wanted to test them within a game rather than just an environment. We presented this game as well as the same game with just static lighting for comparative analysis at the *Interactivity* venue of *Computer Human Interaction Conference* 2005 [23]. Several people played the demo after the researcher explained the premise of the game. All participants who played the game with the lighting system were also invited to play the game without the lighting system (i.e., using a static lighting design); this is important to clearly identify the difference in the experience.

Through observation and interaction with the participants, many interesting observations were made. Participants were excited about the system and voluntarily came to discuss their experience with the author after their play session. An interesting result was that many non-FPS players loved the game and the effect of the lighting. Some commented that it was beautiful and esthetically pleasing to play with the lighting changes than with just static lighting. Some commented that they saw lighting as a method for portraying game information, which was unique in their experience.

Several FPS gamers played the two versions of the game. Some commented that the lighting gave them too much information and that impeded their game play, i.e., made the game too easy. Several others noted some disturbance by the lighting. One explanation was that many FPS players try to emotionally detach themselves from the game, but the lighting effects subconsciously attempt to draw the players in by manipulating the projected tension. In addition, some of them commented that this effect made them feel as if they are not in control. Perhaps this result confirms the success of the patterns in projecting tension, but alludes to the fact that the use of these patterns may need more study for different game genre. Further investigations are needed within this direction to confirm the effect on lighting on emotions and validate the patterns discussed within a game.

6 Conclusion

In this chapter, we discussed several lighting systems: ELE, ALVA, and TDELE. ELE was discussed first as it was the base for the other two lighting systems. ALVA and TDELE were developed to explore the esthetic value of procedural lighting in evoking tension and affect (TDELE) and manipulate visual attention (ALVA) in real-time. Experimental results show that using ALVA was an attractive alternative for casual and non-gamers, i.e., non-FPS gamers, as it can adapt to their visual abilities. An industry person from Microsoft commented that the system can be used to train non-FPS gamers to gradually become better within FPS games if we can dynamically tune out the lighting emphasis on visual attention. This may be a good use of such a system, and it can be a great tool to make FPS games more accessible. However, the experiments discussed are limited. For example, the number of subjects is really small. Also, we did not measure the impact of reaction time and how that may have impacted the data or results collected. Nevertheless, the results are promising and reinforce that this area of research is important, but requires more research to establish its utility within a game. Experiments with TDELE also show some promise. The results show clearly that participants all went through the same reaction, given the environments they were asked to navigate through. Yet again, more experiments are needed to explore this direction further.

As the industry starts to expand its market, adaptable game systems will become increasingly important. As this shift starts to occur, there will be a need for adaptive and procedural systems. One cannot think of adaptive game systems without adaptive esthetics. As argued before, visual esthetics is very important to game development. Therefore, when one embarks on the road of building procedural games, one must start exploring and integrating procedural visual esthetics systems, such as camera and lighting. Therefore, we see opportunities in the use of the systems and theories discussed in this chapter in next generation games. However, the research is still in its infancy. These systems are still small explorations in a big design space that needs more work to uncover. Work is also needed to establish more theoretical models around these results.

References

1. Almeida, P.: Identifying low-level visual patterns that stimulate emotions and moods in movies and video games. Master's thesis, Penn State University (2005)
2. Alton, J.: Painting with Light. Berkeley: University of California Press (1995)
3. Baid, C.E.: Drama. W. W. Norton & Company (1973)
4. Bates, J.: The role of emotion in believable agents. Communications of the ACM **37**, 122–125 (1992)
5. Bates, J., Loyall, B., Reilly, S.: An architecture for action, emotion, and social behavior. Tech. Rep. CMU-CS-92-144, Carnegie Mellon University, Pittsburgh (1992)
6. Birn, J.: Digital Lighting & Rendering. New Riders Press (2000)
7. Blinn, J.: Simulation of wrinkled surfaces. In: ACM SIGGRAPH, pp. 286–292 (1978)

8. Block, B.: The Visual Story: Seeing the Structure of Film, TV, and New Media. Focal Press, New York (2001)
9. Bordwell, D., Thompson, K.: Film Art: An Introduction, 6th edition edn. Mc Graw Hill (2001)
10. Brown, B.: Motion Picture and Video Lighting. Focal Press, Boston (1996)
11. Calahan, S.: Storytelling through lighting: a computer graphics perspective. In: Siggraph Course Notes (1996)
12. Campbell, D.: Technical Theatre for Non-technical People. Allworth Press (1999)
13. Carson, D.: Environmental storytelling: Creating immersive 3d worlds using lessons learned from the theme park industry. Gamasutra (2000)
14. Cassell, J., Pelachaud, C., Badler, N., Steedman, M., Achorn, B., Becket, T., Douville, B., Prevost, S., Stone, M.: Animation conversation: rule-based generation of facial expression gesture and spoken intonation for multile conversational agents. In: SIGGRAPH Computer Graphics (1994)
15. Chen, H., Liu, X.: Lighting and material of halo 3. In: ACM SIGGRAPH, p. 22 (2008)
16. Cheshire, D.: The Book of Movie Photography. Alfred Knopf, Inc. (1989)
17. Crowther, B.: Film Noir: Reflections in a Dark Mirror. Columbus, London (1989)
18. El-Nasr, M.S.: Automatic expressive lighting for interactive scenes. Tech. rep., Computer Science Evanston, IL: Northwestern University (2003)
19. El-Nasr, M.S.: Intelligent lighting for game environments. Journal of Game Development **1** (2005)
20. El-Nasr, M.S., Almedia, P.: Projecting tension in virtual environments through lighting. In: ACM SIGCHI International Conference on Advances in Computer Entertainment Technology (2006)
21. El-Nasr, M.S., Horswill, I.: Automating lighting design for interactive entertainment. ACM Computers in Entertainment **2** (2004)
22. El-Nasr, M.S., Ioerger, T., Yen, J.: FLAME – fuzzy logic adaptive model of emotions. Autonomous Agents and Multi-Agent Systems **3**, 219–257 (2000)
23. El-Nasr, M.S., Miron, K., Zupko, J.: Intelligent lighting for a better gaming experience. In: Computer Human Interaction (2005)
24. El-Nasr, M.S., Morie, J., Drachen, A.: A scientific look at the design of aesthetically and emotionally engaging interactive entertainment experiences. In: Gökçay, D., and Yildirim, D. (eds.), Affective Computing and Interaction: Psychological, Cognitive and Neuroscientific Perspectives, pp. 281–307. IGI Global (2011)
25. Foss, B.: Filmmaking: Narrative & Structural Techniques. Los Angeles: Silman-James Press (1992)
26. Gillette, J.M.: Designing with Light, 3rd edition edn. Mountain View (1998)
27. Green, R.: Spherical Harmonic Lighting: The Gritty Details. Sony Computer Entertainment America (2003)
28. Greuter, S., Parker, J., Stewart, N., Leach, G.: Undiscovered worlds – towards a framework for real-time procedural world generation. In: International Digital Arts and Culture Conference. Melbourne, Australia (2003)
29. Handprint-Media: Light and the eye. Web Page (2001)
30. Katra, E., Wooten, B.: Preceived lightness/darkness and warmth/coolness in chromatic experience. Master's thesis, Unpublished MA thesis, Brown University (1995)
31. Koonce, R.: GPU Gems 3, chap. Deferred Shading in Tabula Rasa, pp. 429–457. Addison-Wesley (2007)
32. Lagrange, M., Marchand, S., Raspaud, M., Rault, J.: Enhanced partial tracking using linear prediction. In: International Conference on Digital Audio Effects (DAFx), pp. 141–146 (2003)
33. Loyall, B.: Believable agents. Tech. rep., Computer Science Department Pittsburgh: Carnegie Mellon University (1997)
34. Mateas, M., Stern, A.: A behavior language for story-based believable agents. IEEE Intelligent Systems **17**, 39–47 (2002)
35. McTaggart, G.: Half-Life 2 / Valve Source Shading. Valve Corporation (2004)
36. Millerson, G.: The Technique of Lighting for Telivision and Film, 3rd edition edn. Focus Press, Oxford (1991)

37. Mittring, M.: Finding next gen – cryengine 2. In: ACM SIGGRAPH, pp. 97–121. San Diego, California (2007)
38. Morie, K.I.J., Valanejad, K., Sadek, R., Miraglia, D., Milam, D.: Emotionally evocative environments for training. In: 23rd Army Science Conference (2002)
39. Mudur, S.P., Pattanaik, S.N.: Computation of global illumination by monte carlo simulation of the particle model of light. In: Third Eurographics Workshop on Rendering (1992)
40. Nelson, M.J., Mateas, M.: Towards automated game design. In: Procedings of the 10th Congress of the Italian Association for Artificial Intelligence (AIIA 2007), pp. 626–637. Rome, Italy (2007)
41. Niedenthal, S.: What we talk about when we talk about game aesthetics. In: DIGRA (2009)
42. Perlin, K.: Building virtual actors who can really act. In: International Confernece on Virtual Storytelling (2003)
43. Platoniotis, K.N., Venetsanopoulus, A.N.: Color Image Processing and Applications. Springer (2001)
44. Prachyabrued, M., Roden, T.E., Benton, R.G.: Procedural generation of stylized 2d maps. In: International Conference on Advances in Computer Entertainment Technology. Salzburg, Australia (1997)
45. Smelik, R.M., Tutenet, T., de Kraker, K.J., Bidarra, R.: A proposal for a procedural terrain modelling framework. In: EGVE Symposium (2008)
46. Sousa, T.: GPU Gems 3, chap. Chapter 16. Vegetation Procedural Animation and Shading in Crysis. Addison-Wesley (2007)
47. Swartout, H., Gratch, J., Kyriakakis, L., et. al.: Toward the holodeck: Integrating graphics, sound, character and story. In: Fifth International Conference on Autonomous Agents, pp. 409–416 (2001)
48. Treisman: The preception of features and objects. In: Attention: Selection, awareness and control, pp. 5–35. Clarendon Press University (1993)
49. Treisman, A.M., Gelade, G.: A feature-integration theory of attention. Cognitive Psychology **12**, 97–136 (1980)
50. Yu, Y., Debevec, P., Malik, J., Hawkins, T.: Inverse global illumination: recovering reflectance models of real scenes from photographs. In: International Conference on Computer Graphics and Interactive Techniques (1999)
51. Zangwill, N.: Aesthetic judgment. In: Stanford Encyclopedia of Philosophy (2003)
52. Zupko, J.: A system for automated interactive lighting (SAIL). Tech. rep., State College, Pennsylvania: The Pennsylvania State University (2009)

Index

P.A. González-Calero and M.A. Gómez-Martín (eds.), *Artificial Intelligence for Computer Games*, DOI 10.1007/978-1-4419-8188-2,

U

V

W